HUBBLE
REVISITED

For Tom

Christmas 1999

on your long distance
arm chair convalesence travels

much love,
Jane

HUBBLE
REVISITED

New Images from the Discovery Machine

DANIEL FISCHER   HILMAR DUERBECK

Translated by
Helmut Jenkner

Foreword by Steven A. Hawley

COPERNICUS
AN IMPRINT OF SPRINGER-VERLAG

Originally published as *Das Hubble-Universum: Neue Bilder und Erkenntnisse*, © 1998 by Birkäuser Verlag, Basel, Switzerland.

Published in the United States by Copernicus, an imprint of Springer-Verlag New York, Inc.

Copernicus
Springer-Verlag New York, Inc.
175 Fifth Avenue
New York, NY 10010

Library of Congress Cataloging-in-Publication Data

Fischer, Daniel.
    Hubble revisited: new images from the discovery machine /
Daniel Fischer, Hilmar Duerbeck ; [translated by Helmut Jenkner].
      p. cm.
    "Copernicus imprint."
    Includes bibliographical references and index.
    ISBN 0-387-98551-4 (alk. paper)
    1. Hubble Space Telescope (Spacecraft) 2. Outer
space—Exploration. I. Duerbeck, Hilmar, W., 1948–
    II. Title.
QB500.268.F575   1998
522′.2919—dc21   98-29991

Manufactured in Italy.

Printed on acid-free paper.

9 8 7 6 5 4 3 2 1

ISBN 0-387-98551-4   SPIN 10680624

Table of Contents

Foreword   7
Preface   11

Part 1. Telescopes on the Ground and in Space   13

Telescopes in Orbit and New
  Telescopes on Mountaintops   15
The Long Road to Hubble's Launch   23
In Orbit   26
The First Servicing Mission   30
The Second Servicing Mission   33

Part 2. To the Edge of the Universe   41

The Basic Questions of Cosmology   43
A View into the Depths of Space
  and Time: The Hubble Deep Field   46
The Search for Cosmic Numbers   56
Searching for the Building Blocks of
  Galaxies   65
Colliding Galaxies   68
Quasars: Beacons at the Beginning
  of Time   76
Active Galaxies: Nearby
  Mini-Quasars   80

Gravitational Lenses: Hubble's
  Telephoto Lens   82
Cosmic Gamma Ray Bursts   88

Part 3. Stars   95

Stellar Nurseries   97
Giant Stars   112
Globular Clusters, White Dwarfs, and
  Blue Stragglers   118
Still Going Strong: Supernova 1987A   121
Neutron Stars   126
A Festival of Colors and Shapes   128
Old Couples   145

Part 4. Planets   149

Planets around Other Stars   151
At the Edge of the Solar System:
  Pluto and Trans-Neptunian
  Objects   155
Neptune and Uranus   157
Saturn   161
Jupiter's Aurorae   166
Jupiter's Moons   169
Asteroids   172
Comets   175
Mars   177

Part 5. Hubble's Future and Its Successors **183**

Hubble's Second Decade   185
The Next Two Shuttle Visits   188
Hubble's Legacy: The Data Archive   190
Hubble's Successors   192
Europe as Partner   198

Part 6. Appendix **201**

Glossary   203
Want to See More? (World Wide
    Web Addresses)   209
Further Reading   210
Index   213

Foreword

As a small boy, I dreamed that one day there would be telescopes in space. That seemed to me a logical progression given the growth of light pollution and the scientific desire to expand the coverage of accessible wavelengths. I even imagined that there would be a need for trained astronomers to go into space to conduct observations with orbiting telescopes. Perhaps one day I would be lucky enough to be one of them.

As I grew older I came to see this dream as unlikely. There would be telescopes in space, indeed, but the data would be relayed to Earth digitally. There would be no need for astronomers in space.

Still, to be an astronomer at this time in history is a blessing. Scientists are developing an unprecedented understanding of our universe, its components and processes, and we are beginning to answer the age-old questions: How old is the universe? How big is it? What is its ultimate fate? Are there other planets? Where do galaxies come from? How are stars born, and how do they die?

As it turned out, my dreams did come true. Although I did not observe the universe through the Hubble Space Telescope, as I once imagined as a child, I did become an astronomer in space and was one of the very few humans lucky enough to see Hubble up close in orbit. As a member of both the original deployment mission crew, in 1990, and the second servicing mission crew, in

1997, I actually experienced this tremendous sight twice.

Hubble is a beautiful sight up close. At a distance of more than 40 miles, it outshines any celestial object except the Sun, the Moon, and the Earth. As the space shuttle *Discovery* approached within several hundred feet of the telescope, I could see the shiny aluminum blankets that provide thermal protection reflecting the light from the Earth's oceans and making the telescope appear bluish instead of silver. When Hubble was latched in *Discovery*'s payload bay, I saw the solar arrays glow with reflected sunlight, appearing as though they had their own source of internal illumination. On the second mission, the side of the telescope that points toward the Sun showed signs of "weathering" from its seven years in orbit. Some of the insulation is cracked and peeling, and the silver surface appears less shiny, as though it had been brushed with steel wool. But these are superficial blemishes—the real beauty of the Hubble Space Telescope is its phenomenal performance.

The original design of the telescope was visionary in that it allowed for on-orbit servicing to mitigate the consequences of the inevitable failure of some components. Even more important, it allowed new technology to be incorporated into the telescope's detectors so that Hubble would remain a state-of-the-art observatory throughout its designed lifetime of 15 years. Regular servicing missions at about three-year intervals have occurred since the initial deployment mission in 1990.

The first servicing mission, in 1993, restored Hubble's performance to its original specifications. That mission, conducted under much public scrutiny, was generally known as the "Hubble repair mission." The second servicing mission, in February 1997, built on the successes of the first. This "Hubble upgrade" mission extended the telescope's view into the infrared and increased the efficiency of data collection by more than an order of magnitude. The scientific benefits of the enhancements from that mission are just beginning to be realized and won't be fully known for many years. Just as important, I believe, was the demonstration that orbiting facilities like Hubble can be maintained and upgraded so that their performance can keep pace with technology development and remain state of the art. The two servicing missions demonstrated our ability both to maintain the telescope and to enhance the scientific return. The missions improved a telescope that was already performing world-class science. The next servicing mission is scheduled for the spring of 2000 and will again perform both maintenance and upgrades. Among the activities currently scheduled are installation of a new set of solar arrays and a new camera called the Advanced Camera for Surveys.

In the few months since the installation of the

Near-Infrared Camera and Multi-Object Spectrometer (NICMOS) and the Space Telescope Imaging Spectrograph (STIS), these second-generation instruments have more than lived up to astronomers' expectations. Hubble has photographed the most luminous star known and a neutron star alone in space, resolved the Mira double system, revealed new insights into black holes, and discovered the farthest galaxy ever detected in the universe. Photographs such as those in this book foretell the amazing images and profound discoveries we can anticipate in the years to come.

Seven years after Hubble's initial deployment, the public continues to be captivated by the images that it has obtained as well as by the scientific interpretations of those images. Hubble's performance will continue to be upgraded by the development of newer instruments and their installation on future servicing missions. As long as we can continue to replace aging components, the facility will be able to perform world-class science for the foreseeable future. Even as scientists

Astronaut Steven A. Hawley observes the activities of two colleagues on the monitor during the second servicing mission to the Hubble Space Telescope in 1997. Hawley was also part of the space shuttle crew that deployed HST in 1990. (NASA)

and engineers consider what the Next Generation Space Telescope should be, the Hubble Space Telescope will continue to reveal many unprecedented details of the wonders of the universe.

Steven A. Hawley, Ph.D.
NASA Astronaut
HST Deployment Mission
HST Second Servicing Mission

Preface

Three years ago, we collected pictures and wrote a book about the Hubble Space Telescope. We aimed to create an illustrative and readable book from the wealth of discoveries, reports, and images available; at the end, we were fairly pleased with our efforts. Developments kept trying to overtake us, but we assembled complete and up-to-date information of interest to the reader in 1995. More or less updated versions of our book in English, French, and Japanese followed the German edition and were received favorably. The wealth of discoveries and new developments in space- and ground-based astronomy made a new edition desirable. But the vast increase in the number of images and amount of knowledge acquired since the first edition, and significant new developments on the ground and in space, made it clear that the simple addition of a few updated sections would not do justice to the new situation.

Therefore, we decided to write a completely new book about Hubble. This book continues the basic theme of the previous one and does not leave out any significant details of Hubble's early years. But its main emphasis is in the present: the Hubble Space Telescope with its improved capabilities. The first servicing mission in 1993 furnished corrective "glasses" for the telescope, and the second mission in February 1997 provided it with completely new "eyes." The improved Hubble Space Telescope explores the depths of the universe to the limits of space and time,

as shown by the Hubble Deep Field. It probes through the dust veils that obscure the birth of stars. And it reveals new aspects of our own solar system: surprise visitors from outer space—comets Hyakutake and Hale-Bopp—inspired the general public as well as astronomers.

Recent years have seen the inauguration of several impressive ground-based optical telescopes, which we discuss. We also describe the current astronomical satellites. In the last chapter, we take a look at the not-so-distant future, when the Hubble Space Telescope will be replaced by astronomy satellites that apply new technologies.

Once again, the Internet made the collaboration between the authors simple and effective, and we are grateful for its existence. For his willingness to provide a foreword to our book, we thank astronaut and astronomer Steve Hawley, who released the telescope into space during the deployment mission in 1990 and who was a mission specialist again on the second servicing mission. We owe our gratitude to a number of individuals; without their interest and contributions, this book could not have been written. H.D. thanks Nino Panagia and Bob Williams for the opportunity to work at the Space Telescope Science Institute for three months during the spring of 1997; there, he was able to witness the second servicing mission and its results firsthand, including the fascinating report of the astronauts after their return from space and the meeting on the Hubble Deep Field and its importance for science. D. F. is grateful to the European Space Agency for the opportunity to participate in the meeting on "Science with the Hubble Space Telescope II" at the end of 1995 and to the American Astronomical Society for the opportunity to take part in their 191st meeting at the beginning of 1998; these meetings brought the central role of Hubble in modern astronomy into clear focus.

We also thank the staff of the Space Telescope Science Institute and all those scientists that were able to obtain much-sought-after observing time on the Hubble Space Telescope. Only their untiring work in preparing the results from the telescope—in calibrating, analyzing, and interpreting the data—makes the Hubble Space Telescope such a success story. Special thanks are due to those who generously made Hubble results available for this book, in part before publication. We particularly thank our translator and helper with many things, Helmut Jenkner. Large parts of the text were written while H. D. stayed at the Universidad Catolica del Norte in Antofagasta, Chile, under the sponsorship of the German Academic Exchange Service—practically in the shadows of the Very Large Telescope of the European Southern Observatory. He expresses his gratitude for several months of hospitality to the staff of the Instituto de Astronomia of the Universidad Catolica del Norte and particularly to its director, Luis Barrera Salas.

Daniel Fischer and Hilmar Duerbeck
Königswinter, Germany; Antofagasta, Chile; and
    Baltimore, Maryland
Spring 1998

Part 1

Telescopes on the Ground and in Space

Telescopes in Orbit and New Telescopes on Mountaintops

Immediately after World War II, when the first plans for an astronomical telescope in space were taking shape, the world in general—and the world of astronomers in particular—was a different one. The construction on Palomar Mountain in California of a giant telescope designed by George Ellery Hale, with a mirror diameter of 5 meters (200 inches), had taken several years and posed significant technical problems. It turned out to be the only large telescope in astronomy for several decades. It was a long and tedious process to top this great technological achievement: a Russian 6-meter telescope approximated the quality of its predecessor only much later and after a number of necessary improvements, and the jump to 8-meter-class telescopes did not occur until the beginning of the 1990s. Not all the great questions of astronomy could be answered from the ground, however, as was already clear to several visionaries during the first half of the twentieth century—even before the first artificial satellite had reached orbit.

When the first *Sputnik*s, *Explorer*s, and *Vanguard*s led the way into space from 1957 onward, they may have owed their existence mainly to political motivations. But from the very beginning, scientists tried to exploit the new possibilities. For the first few years, they focused on the space close to Earth, but by the 1960s, the first astronomical observatories were launched into orbit. These very small observatories were only distant cousins of the optical observatories on the ground. The limited lift capabilities of the first rockets and the lack of powerful electronic detectors made it unrealistic to launch telescopes to observe the same visible light that could be gathered and analyzed from ground-based observatories much more easily. When this situation later changed, the advantages of space-based over ground-based astronomy could be fully exploited.

There are three main factors that make a space telescope superior to an observatory on the ground, even in the best possible location:

- A telescope in space is not limited by air turbulence. At a ground-based observatory, the turbulence of the Earth's atmosphere renders images of the sky blurred and hazy; even the best methods of applied optics and image processing cannot fully compensate for this effect.
- From space, the sky is black and cloudless. Even the best observatory locations on the ground suffer from the natural faint glow of the atmosphere (not to mention close artificial sources of light) and from changes in transparency.
- Finally, the absence of an atmosphere allows a space telescope to observe not only in visible light but also in those spectral regions where the radiation reaches the ground either not at all or strongly weakened.

The first satellites designed for astronomy carried special detectors for the various types of radiation that cannot be observed from the ground, particularly

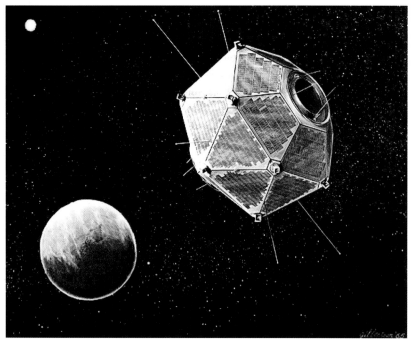

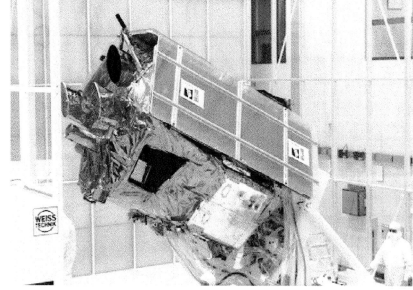

A gallery of telescopes in orbit. Top left: A *Vela* satellite, originally designed to detect nuclear tests on Earth; four of these satellites observed a total of 73 gamma ray bursts from space between 1969 and 1979. (LANL). Top right: The Compton Gamma Ray Observatory during its deployment by the space shuttle in 1991. (NASA). Bottom left: The Infrared Space Observatory; from 1995 to 1998, it explored longer wavelength regions than those accessible to the Hubble Space Telescope. (ESA). Bottom right: The Italian-Dutch x-ray satellite Beppo-SAX; it helped to solve the riddle of the cosmic gamma ray bursts in 1997. (ASI)

the far ultraviolet and x-ray domains. The Earth's atmosphere shields us from UV and x-ray radiation, which has shorter wavelengths than that of visible light. Without this atmospheric shield, these types of radiation would severely affect our health. But this area of the electromagnetic spectrum is very well suited to the exploration of exotic objects in the universe: hot stars, glowing disks of matter, even black holes. Instruments capable of detecting these types of radiation had been developed by nuclear physicists, and they could now be adopted for use in space. In the ultraviolet and x-ray range, these early satellite telescopes could see the sky only with a resolution comparable to that of a near-sighted person without glasses in visible light. Nevertheless, they opened new windows onto the universe, much as radio astronomy had done from the ground several decades earlier.

The early space telescopes led to the unexpected discovery of some of the more exotic objects in the universe. Various satellites searched for the characteristic radiation from nuclear bombs detonated on Earth or behind the Moon in violation of test ban treaties. The *Vela* satellites of the U.S. Air Force, for instance, looked for bursts of gamma radiation, which is even more energetic (or "harder") and has shorter wavelengths than x rays. And they did indeed find such gamma ray bursts: they were more frequent and weaker than expected from nuclear explosions, and did not originate on Earth or behind the Moon, but in the depths of space. These scientific discoveries

remained hidden from the public for years. The first publication on this subject did not appear until 1973, and the source of these gamma ray bursts remained a mystery for decades. It was not until 1997 that the concerted observations of several telescopes and satellites, the Hubble Space Telescope (HST) among them, finally provided some insight into this phenomenon—as we will see later.

The Earth's atmosphere is also largely opaque to wavelengths longer than that of visible light. It absorbs not only the incoming infrared radiation from space but also the heat emanating from Earth. Our atmosphere acts as a blanket, insulating the warm Earth from the extreme cold of outer space. Because very young and very old stars are often comparatively cool and the dust that may envelop them radiates at these longer wavelengths, we must explore the infrared region of the spectrum to learn about the birth and death of a star. To do this, infrared telescopes have been built on high mountains or installed in high-flying airplanes and balloons, where the absorption of the atmosphere is significantly reduced. And as soon as the appropriate detector technology became available for use in space, special infrared satellites were launched into orbit.

The telescopes carried on these satellites—among them the Infrared Astronomy Satellite (IRAS), Cosmic Background Explorer (COBE), and Infrared Space Observatory (ISO)—are relatively small because they must be cooled to very low temperatures. They are located inside large dewars ("Thermos® bot-

Giant telescopes on Mauna Kea, Hawaii; the shadow of this dormant volcano projects onto the cloud layer shortly after sunrise. The domes of the Japanese 8-meter telescope Subaru and of the two 10-meter Keck telescopes are located in the foreground. (National Astronomical Observatory of Japan)

tles"), and their mirror diameters of less than a meter are about the same as good amateur telescopes on the ground. Because these infrared telescopes in space observe at much longer wavelengths (typically 3 to 200 micrometers, or microns), their resolution is only as good as that of good binoculars in the visible light range!

The Hubble Space Telescope, finally launched in 1990 after two decades of preparation, contains the largest mirror of all telescopes in space for scientific applications. For that reason, it was desired to design HST for observation in the near infrared, as well as in the visible and ultraviolet ranges of the spectrum. The required detectors did not become available until after HST's launch, but the first infrared instrument was installed on HST in 1997, as described later in this chapter. HST's telescope itself cannot be cooled, however, so it is still impossible to observe at the longer wavelength domain of the infrared region. The radiation coming from the "hot" telescope itself would drown out the signals from faint celestial objects.

The Next Generation of Ground-Based Telescopes

Astronomical objects reveal their true nature only in the analysis of data from different observations, applying various telescopes on the ground and in space with instruments that cover different wavelength ranges. Later, we will encounter many results and answers to some of the fundamental questions in astronomy that are based on the combination of observations with Hubble and with other instruments. Even as the Hubble Space Telescope orbits our planet, new giant telescopes on the ground are far surpassing its light-gathering power. We will survey some of them in this section. The multitude and magnitude of these new ground-based observatories are partly due to the revolutionary success of Hubble. On the other hand, they provide significant help in understanding some of its results and in preparing for later Hubble observations. We will not attempt to enumerate the pros and cons of ground-based versus space-based telescopes here—it will become obvious in the remainder of this book that both are required for progress in astronomy today.

The W. M. Keck Observatory

The Keck Observatory consists of two 10-meter telescopes operated by the California Institute of Technology, several astronomical institutions of the University of California, and NASA. It is located atop Mauna Kea, a dormant volcano on the island of Hawaii, at an altitude of 14,000 feet. Each telescope weighs in at 300 metric tons. Each primary mirror consist of 36 hexagonal 1.8-meter segments that are actively controlled. Twice per second, each mirror segment is brought to its optimal shape to achieve a near-perfect focus with a precision of 4 nanometers (a nanometer is one-billionth of a meter—about one-thousandth of the diameter of a human hair). Spectrographs and cameras for visible light and for infrared wavelengths are used on these telescopes. Keck I began operating in May 1993 and Keck II in October 1996. The cost of more than $140 million was borne by the W. M. Keck Foundation.

The Subaru Telescope (Japan National Large Telescope, JNLT)

This 9.3-meter telescope, also located atop Mauna Kea, Hawaii, is operated by the National Astronomical Observatory in Tokyo. Construction began in 1991, and "first light" is expected in the summer of 1998. The telescope will commence full scientific operation in the year 2000.

The Very Large Telescope (VLT)

The VLT consists of four individual telescopes of 8.2-meter diameter. These telescopes will be able to operate individually or together; the four telescopes operating together, corresponding to a single mirror with a diameter of 16 meters, will have the largest light-gathering power of any telescope under development. The VLT is being constructed by the European Southern Observatory (ESO); the first of the four telescopes began operating during the summer of 1998. The VLT is located at an altitude of 8600 feet on top of Paranal Mountain in the Chilean Andes, about 80 miles south of the town of Antofagasta. Although the Pacific Ocean is only about 8 miles away, the observatory sits high above the clouds in one of the driest locations on Earth.

The Gemini Project

This international project plans to construct two 8-meter telescopes. One will be located on Mauna Kea, Hawaii; the other on Cerro Pachon, near La Serena, Chile. Participants in the project include institutions in the United States, the United Kingdom, Canada, Chile, Argentina, and Brazil.

The Magellan Project

Two 6.5-meter telescopes are planned for Las Campanas Mountain in Chile's Atacama Desert. The Carnegie Institution of Washington and a number of U.S. universities participate in this project. The first

Europe's response: the Very Large Telescope (VLT) array on Cerro Paranal in Chile is nearing completion. Four VLT Unit Telescopes, each with a mirror 8.2 meters in diameter, will take advantage of the excellent climate near the Chilean coast (see image from space). Test observations with Unit Telescope 1 (left, and in the large image during completion in Italy) started in May 1998. (ESO and NASA)

telescope is scheduled to begin operation in the year 2000.

The Hobby-Eberly Telescope (HET)

This telescope, located at the McDonald Observatory in Fort Davis, Texas, consists of a segmented main mirror with an effective diameter of 9 meters. It is operated by a consortium of universities, including the University of Texas at Austin; Pennsylvania State University; the University of California at Stanford; and the University Observatories of Munich and Göttingen, Germany. It is a "transit telescope," meaning that it can only move around a horizontal axis; however, it can follow celestial objects over a period of time with a movable secondary mirror. This makes about 70 percent of the available sky accessible for observation. The telescope is intended primarily for spectroscopic observations. The unusual design led to a significant reduction of the telescope's cost, amounting to "only" $14 million. Observations began in 1997.

The Large Binocular Telescope (LBT)

The LBT will be constructed on Mount Graham, Arizona, at an altitude of 10,600 feet. Participants are the University of Arizona; Ohio State University; the Arcetri Astrophysical Observatory in Florence, Italy; the German Max Planck Institutes for Astronomy in Heidelberg, for Extraterrestrial Physics in Munich, and for Radio Astronomy in Bonn; and the Astrophysical Observatory of Potsdam, Germany. The optical assembly consists of two parallel telescopes with 8.4-meter primary mirrors only 14.4 meters apart. The use of interferometry will allow an image resolution comparable to that of a single 22-meter mirror. The total project cost is estimated to be about $140 million. Scientific operation is planned to begin during 2003 and 2004. Scientific goals for this telescope include cosmology, the formation of stars and planets, and direct imaging of planets around nearby stars.

Telescopes in Space

Results from the Hubble Space Telescope are frequently combined with observations from other astronomical satellites, in ways similar to the complementary role of ground-based observatories. Over the last several years, the various space agencies have been active. New astronomical satellites have been and are being developed and launched. We will introduce a number of them in this section. In general, these astronomical satellites observe in wavelength ranges that are inaccessible from Earth, either in the infrared region or in the x- and gamma ray regime. The new space observatories are not intended to compete with the Hubble Space Telescope but to cover the broad electromagnetic spectrum better.

Two infrared satellites can observe at long wavelengths. The first one, the European Infrared Space Observatory (ISO), was launched in November 1995. It features a 0.6-meter mirror and a camera, two spectrometers, and a polarimeter that are capable of analyzing infrared radiation with wavelengths between 3 and 200 microns. ISO was operational until April 1998, when its 2140 liters of liquid helium for cooling the telescope and detectors were exhausted. After another month of technical experiments and a few scientific observations at shorter wavelengths, ISO was finally switched off on May 16. In several years, it will burn up in the atmosphere—but its legacy will remain.

The U.S. Space Infrared Telescope Facility (SIRTF), currently under development, consists of a 0.86-meter mirror and three scientific instruments. It will provide images and spectra in the wavelength range from 3 to 180 microns with significantly better performance than ISO. Its launch is planned for 2002.

Even farther into the future, in about 2007, lies the planned launch of the Far Infrared and Submillimeter Space Telescope (FIRST) of the European Space Agency (ESA). Its 3-meter mirror and infrared and radio detectors will detect radiation with wavelengths between 90 and 900 microns. Molecules in space and protostars emit such radiation. FIRST will also provide the opportunity to look at the "dark ages" of the universe, when the first generation of stars and galaxies began to form. Because the universe is expanding, the radiation from these objects is strongly shifted toward longer wavelengths, into the far infrared. These objects can be observed only by large infrared telescopes such as FIRST or the even larger successor to the Hubble Space Telescope, the Next Generation Space Telescope (NGST). We will present more details of the NGST project in the last part of the book.

New satellites also have been launched or are under development to observe the other end of the electromagnetic spectrum. Of special interest are the German x-ray Roentgen Satellite (ROSAT), launched in 1990, and the Italian-Dutch x-ray satellite Beppo-SAX (Satellite per Astronomia X "Beppo," in honor of Italian physicist Giuseppe "Beppo" Occhialini), launched on April 30, 1996; the latter provided important information for the localization of the mysterious gamma ray bursters. In addition to a flotilla of highly specialized small satellites, the U.S. and European space agencies are planning to launch two next-generation x-ray satellites of significant size, beginning in late 1998: the NASA Advanced X-Ray Astrophysics Facility (AXAF) and the ESA X-Ray Multi-Mirror Mission (XMM). The two satellites complement each other in their abilities to obtain detailed x-ray images and deep spectra. Russia planned to develop sophisticated x-ray and gamma ray space observatories, but their completion is increasingly uncertain because of an acute lack of funds.

In the coming decades, a number of technologies will probe cosmic radiation over a wide wavelength range, better and more completely than ever before. But it is hoped that the Hubble Space Telescope will continue to deliver unique views from orbit during its second decade of operation. The road to this success, however, has been long and difficult.

The Long Road to Hubble's Launch

In 1946, the eminent American astronomer Lyman Spitzer envisioned a telescope orbiting Earth. Although an even earlier idea came from the German space flight pioneer Hermann Oberth, his concept concentrated on practical applications: the telescope was to be used by future astronauts to warn them of asteroids crossing their path. Spitzer, however, was fully focused on basic research. "The most important contribution of such a radically new and powerful instrument" he wrote then, "would not be to supplement our current ideas about the universe in which we live, but to uncover new, hitherto unimaginable problems." But in 1946, the 5-meter telescope on Palomar Mountain was just being completed, and nobody paid attention to Spitzer's idea.

It was not until the launch of *Sputnik* in 1957, more than a decade later, that proposals for a telescope in space began to be taken seriously. In 1969, Spitzer proposed that a 3-meter-diameter telescope be placed in orbit. After the abrupt end of the *Apollo* program, NASA was looking for a project as spectacular as the lunar landings. The idea of a Large Space Telescope (LST) emerged, but the required funds were not yet available. In 1972, NASA began development of the first reusable space transport system, the space shuttle. LST was linked to the shuttle concept from the very beginning. Not only would the telescope be launched into orbit with the space shuttle, it would be serviced at regular intervals. Various components, worn-out parts, and aging scientific instruments could be exchanged by astronauts, which would allow the space telescope to remain in operation for decades, similar to ground-based observatories.

The first attempt to finance the development failed; Congress did not like the expected $400–$500 million price tag, and the project was postponed. But Lyman Spitzer did not give up. He rallied the American astronomers into a unified group in favor of a space telescope. The first funds for studies were released in 1975, but President Ford called for austerity and the substantial participation of other nations. From 1977 onward, the project began to take shape, even if the telescope had been reduced in size (the mirror diameter was decreased to 2.4 meters). The Lockheed Missiles and Space Corporation in California became the main contractor, having made a name for itself building powerful reconnaissance satellites. The new space telescope was supposed to profit from this technology. The optics, the heart of the satellite, were to be built by Perkin-Elmer in Connecticut.

European scientists were very interested in a space telescope, and there had been contacts between NASA and ESA since 1973. The agencies agreed that ESA would provide the Faint Object Camera (FOC), one of the first five scientific instruments; the technology for such a camera was available at the University College in London. In addition, ESA would provide scientists and engineers to work at the Space Telescope Science Institute in Baltimore, Maryland; establish a European

Lyman Spitzer: The Father of the Hubble Space Telescope

Lyman Spitzer (1914–1997) served as director of the Observatory of Princeton University from 1947 through 1979. He remained at Princeton and worked in his office until the day before his death. In addition to topics in plasma physics and nuclear fusion, Spitzer concentrated on problems in the areas of the interstellar medium, stellar dynamics, and space astronomy.

In 1946, more than a decade before the launch of the first artificial satellite, he proposed in a study for the Rand Corporation that small and large telescopes be developed for use in space. These telescopes would be immune to the problems of image motion and absorption of radiation induced by the atmosphere, and they would not suffer from deformation of their structure by gravity. Small telescopes would glean information about the chemical composition of stars and interstellar gas and dust from the ultraviolet spectra of stars. Large telescopes—Spitzer envisioned mirror diameters between 5 and 15 meters—would determine the structure of globular clusters and galaxies and measure distances in the universe through the analysis of individual stars in distant galaxies.

His first dream became reality in 1972. Under his leadership, the astronomical satellite

American astronomer Lyman Spitzer provided the impetus for the development of a large optical telescope in space and remained involved in the Hubble project until his death in 1997. (Princeton University)

Copernicus, with a main mirror 80 centimeters (32 inches) in diameter, was developed and launched in that year. It would significantly improve our knowledge about the physics and chemical composition of interstellar matter. His second dream was fulfilled with the launch of the Hubble Space Telescope, for which he provided crucial conceptual and practical contributions.

Edwin Hubble at the guiding unit of the Palomar Mountain Observatory's Schmidt telescope in about 1950. The space telescope bearing his name explores the great questions of cosmology that he posed.

development and construction of the Hubble Space Telescope. The final cost of $2 billion was about four times higher than planned. The original launch date of 1983 had to be postponed several times. And when everything was almost ready, the *Challenger* tragedy on January 28, 1986, shut down the shuttle program for two and a half years. The Hubble Space Telescope was forced to spend four years in the dust-free assembly building at Lockheed, until it was finally flown to Cape Canaveral by a large transport plane and installed in the payload bay of space shuttle *Discovery.*

After a final brief delay of 14 days resulting from problems with the auxiliary power units operating the orbiter's hydraulics, *Discovery* lifted off on April 24, 1990, at 8:33:59 A.M. EDT, with "the Hubble Space Telescope, our new window on the universe" on board (as the NASA launch commentator called it in an undoubtedly well-rehearsed phrase). Eight minutes later, the space shuttle arrived at an altitude of more than 600 kilometers, and the astronauts began preparations to release Hubble. Rarely have people held such great expectations at the launch of a satellite—and rarely have they been, at least in the first years, so bitterly disappointed.

Coordinating Facility in Garching near Munich, Germany; and deliver the solar panels for the satellite. In return, astronomers from ESA member countries would receive a minimum of 15 percent of the observing time. A memorandum of understanding including these points was established between NASA and ESA in October 1977.

Financial and technical problems plagued the

In Orbit

Several hours after launch, *Discovery* circled the Earth at an altitude of 613 to 615 kilometers. Every kilometer in altitude gained before the Space Telescope was deployed would prolong its life in orbit—because even at this altitude, there is enough residual atmosphere to slow down such a huge satellite and cause it to descend gradually. Five hours after launch, the satellite transmitted its first signals, but *Discovery* was still providing power. One day later, the "umbilical" cord connecting Hubble to the space shuttle was disconnected. Only eight hours remained to deploy the 12-meter-long solar panels, to prevent the telescope's batteries from draining too much. For launch, the solar panels with their 48,760 individual cells were rolled up in canisters and stored on the sides of the telescope. Fully unfurled, they generate about 4.4 kilowatts of electrical power.

On the morning of the second day, astronaut and astronomer Steve Hawley used the Canadian remote manipulator arm to lift the telescope out of the payload bay. The communications antennae were deployed and the solar panels unfurled. Finally, the entire telescope was released at 3:38 P.M. EDT. *Discovery* moved away very slowly, to avoid last-minute contamination of the satellite by the rocket thrusters, which might ruin Hubble's painstakingly maintained cleanliness. For two more days, the astronauts remained ready to capture the satellite in case of emergency and bring it back to the ground. They had trained for all sorts of emergencies—for instance, the huge front cover of the telescope failing

to open. Indeed, opening the cover took several hours because of communication difficulties, and it put the satellite—for the first time, but not the last—into "safe mode." Whenever Hubble's on-board computers sense danger, they make themselves independent of most commands from the ground. They ensure that the solar panels receive enough light and that the telescope points away from the Sun. It can take hours or even days to bring Hubble out of safe mode.

But these first problems, not totally unexpected in such a complex satellite, were soon dwarfed by much more serious developments. Every time Hubble crossed the day-night boundary in its orbit, major disturbances occurred. It quickly became clear that the huge solar panels were being warped and deformed. This sometimes caused Hubble to lose its orientation and compromised its effective use. The finely tuned gyro systems, which were to be used to move Hubble in all three axes, were too weak to counteract the shaking. But Hubble's largest flaw had not been recognized yet. While the flight controllers and programmers struggled with the unruly satellite, Hubble had not been able to take a single image.

On the morning of May 20, 1990, Hubble's main camera, the Wide-Field and Planetary Camera (WFPC), saw "first light." The target was a rather inconspicuous star cluster in the southern sky, NGC 3532. The first, one-second exposure showed one star; the second, 30-second exposure revealed several more. The camera was working and, apparently, so were the optics. But the structure of the stellar image

Hubble's launch. The space shuttle *Discovery* carries the telescope into orbit on April 24, 1990. (NASA)

but that definitely could not be replaced in orbit: the 2.4-meter main mirror.

In a press conference on June 27, NASA put all the cards on the table. The main mirror of the Hubble Space Telescope was too flat by 2 microns — only a small percentage of the thickness of a human hair, but devastating by optical standards. For a number of technical reasons, but mainly to save money, the complete optical system of Hubble had never been tested with an artificial star, and so the flaw had not been discovered. Single tests performed on the main and secondary mirrors individually had led scientists to expect an excellent complete system. The culprit was found quickly, because the testing apparatus for the main mirror still existed at the manufacturer. For a "null test," a small test lens was positioned 1.3 millimeters too far from the mirror. Why? A laser beam was not reflected at the end of a metering rod, but 1.3 millimeters below it, where paint had chipped off a darkened cap. Unfortunately, the important test equipment itself had never been properly tested.

The second half of 1990 became crucial for the fate of the HST mission: it was even possible that the entire project might be canceled. After discussion of many different proposals, among them such radical measures as the recovery of Hubble and a complete rebuild of the telescope with the existing second (and correctly shaped) main mirror, an elegant solution was finally found: COSTAR (Corrective Optics Space Telescope Axial Replacement). One of Hubble's scientific instruments (the High-Speed

looked strange: in addition to a central bright peak, it showed a blurred disk more than an arc second in diameter, which for the brightest stars revealed a structure reminiscent of tentacles or spider legs.

This problem simply refused to disappear, whatever Hubble's controllers did to correct it. Each day it became more obvious that something had gone wrong and that the entire project was in jeopardy. When the first images of the European Faint Object Camera, taken on June 17, showed the same strange structure, what some of the Hubble optics experts had feared became certainty: the defect was not in WFPC, but in Hubble's main optical system! And the culprit was the component that had been the pride of the project

Photometer, or HSP) would be replaced by a clever mechanism that would place tiny mirrors into the light path of the optics. These mirrors would correct the unfortunate error in the shape of the main mirror and deliver sharp images of the sky to the remaining scientific instruments.

This trick was possible because although the main mirror's shape was badly wrong, it was "perfectly wrong": its shape followed the prescribed form perfectly, only the prescription had been in error. The "tentacles" or "spider legs" that resulted from pointlike stars in Hubble's optics could be exactly reproduced mathematically. And so it was possible to grind additional mirrors with the exactly opposite effect to correct the defect. Among dozens of quite daring concepts to correct the main mirror's flaw, this approach was the most convincing. In 1990, COSTAR was authorized and development began at Ball Aerospace; the successful completion of COSTAR in record time would later lead to several more Hubble-related orders for this company.

Although the specter of cancellation had threatened the Hubble project during the summer of 1990, it now seemed feasible to achieve an almost complete restoration of HST's capabilities within three years. In addition, Hubble did not have to circle Earth idly until it received its corrective "glasses" as part of a servicing mission. After the overly optimistic expectations before launch, and then the public presentation of a total disaster, the situation stabilized. The two spectrographs aboard the Hub-

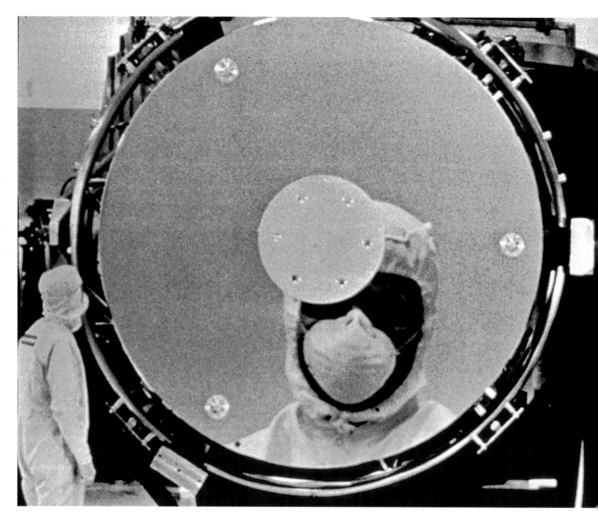

ble Space Telescope were able to begin their work, only marginally affected by the optics flaw. Observations of bright planet surfaces and of similarly bright objects could still be done quite successfully, and computer programs to "sharpen" the Hubble images were developed.

The characteristics of the optical flaw, a spherical aberration, led to images of stars with a large blurred area but with a sharp peak in the center. Unfortunately, only 10 percent of the light reached

Hubble's 2.4-meter main mirror during tests at Perkin-Elmer in the summer of 1984. (NASA)

this central sharp part of the image, but for a number of celestial objects—for instance, star clusters—it was possible to model the blurred halo mathematically and subtract it from the image. This procedure led to an image just as sharp as would have been obtained by a perfect optical system. Only one significant difference remained: faint stars and dim extended objects such as nebulae could not be observed very efficiently, and Hubble's expected sensitivity was reduced, at least for the moment. A number of ambitious observing programs had to be postponed until after the installation of COSTAR, but the majority of projects were still carried out successfully.

At the beginning of 1991, the first research articles based on Hubble data appeared in the scientific literature. The new results included detailed observations of the newly formed giant white cloud on Saturn, the protoplanetary disk around the star Beta Pictoris, the gas ring around the supernova of 1987 in the Large Magellanic Cloud, and the "Einstein Cross"—the image of a quasar multiplied by the gravitational lensing effects of intervening galaxies. If Hubble could produce such impressive results with its restricted vision, what new discoveries would it provide after its repair? The astronomical community was reasonably impressed, but the general public remained skeptical. The troubles had damaged Hubble's reputation too much to trust the current, more optimistic predictions.

The First Servicing Mission

NASA was under tremendous pressure. The first servicing mission (the word "repair" was frowned upon) in 1993 had to be a success, or the future of the entire agency would be in doubt—other than Hubble, it had precious little to show for these years. While Hubble worked on its reduced observing program, valiant efforts continued to correct the problem. Sophisticated mathematical methods of image reconstruction and analysis eliminated the most obvious defects, but attempts to measure the exact brightness of very faint stars above an even fainter background failed. This meant that the planned observation of pulsating stars in other galaxies—one of the key projects of the Hubble Space Telescope (and, much earlier, of its namesake, Edwin P. Hubble)—had to be postponed. Determining a more precise distance scale for the universe would have to wait.

All along, NASA had planned regular servicing missions to Hubble—otherwise the telescope would have been positioned farther away from Earth, rather than in its inefficient and impractical low orbit. Only because the space shuttle had to be able to reach the telescope was it placed in its current orbit, where Earth covers a significant part of the sky and where observations can be affected by Earth's radiation belts. Now the first servicing mission had to be prepared more quickly than planned and needed to include more complex jobs. Two important tasks were the exchange of COSTAR for the High Speed Photometer and replacement of the old WFPC with a new WFCP2 (with its own corrective optics). The old solar panels

also had to be replaced—their vibrations not only disturbed the observations but also posed a threat to the integrity of the entire satellite. The British manufacturer was confident that it had isolated the problem and solved it; the delivery of the new panels was part of the 1977 memorandum of understanding between NASA and ESA.

On December 2, 1993, the space shuttle *Endeavor* lifted off for its eagerly awaited mission. It was a "question of life or death for NASA," as the well-known American astronomer John Bahcall summarized the general sentiment: The mission was carried live not only on NASA TV but also in large parts on the CNN and C-SPAN networks. One of the *Endeavor* crew members was ESA astronaut Claude Nicollier of Switzerland, who operated the remote manipulator arm of the shuttle; only 48 hours after liftoff, he used this arm to capture HST and anchor it safely in the payload bay of the shuttle. On the third day, astronauts Story Musgrave and Jeff Hoffman exchanged two gyro packages that are used to control Hubble's orientation in space. Two (of a total of six) gyros had already failed; these units were expected to need replacement from time to time, however, and they were designed for easy exchange in orbit. The successful completion of the first space walk demonstrated that astronauts could perform maintenance of HST in space—and a "new" satellite began to emerge.

Next, the two old solar panels were commanded to retract in preparation for their exchange. Unfortu-

Astronauts at work. During the first servicing mission in December 1993, all main deficiencies were corrected, giving astronomers the observatory in space they had been dreaming about for several decades. (NASA)

nately, only one panel retracted—the other was so deformed that it was impossible to roll it up again. The next day, during the second extravehicular activity (EVA), astronauts Tom Akers and Kathy Thornton removed the stubborn panel and simply released it into space, where it would orbit Earth for a few more years. The new solar panels were installed without further incident. The first two EVAs ensured the con-

tinued operational safety of the satellite, which had to take precedence over the installation of COSTAR and the new camera. These tasks were planned for the third and fourth space walks.

On the fifth day, Musgrave and Hoffman exchanged the old WFPC for the new (and corrected) WFPC2; on the sixth, Akers and Thornton installed COSTAR and a memory expansion for Hubble's on-board computer. During the fifth EVA — a record — a number of minor repairs were carried out, and the next day, December 10, the newly serviced and fully functional Hubble Space Telescope was lifted out of the payload bay and released back into space. On December 13, at 12:26 A.M. EST, *Endeavor* landed at Cape Canaveral. The entire servicing mission required $674 million, about $100 million of which was a direct result of the optical flaw. But the feasibility of performing complex servicing on a satellite in space had been proved: the four payload specialists had logged a total of $35\frac{1}{2}$ hours of extravehicular activities during this mission.

During the following weeks, necessary adjustments were made to all the new correcting mirrors. On January 13, 1994, a large press conference was held at the Goddard Space Flight Center, where the optical problems had been made public three and a half years before. As prominent political figures and high officials of the space agencies looked on, the energetic Senator Barbara Mikulski stood up, produced two images, and announced: "I am delighted to be able to announce today, after the launch in 1990 and the early disappointments, the trouble with Hubble is over!" Her first picture showed images of single stars, highly magnified, before and after the installation of COSTAR. The hideous halo of light had disappeared, and the new image was perfectly sharp. Even more impressive was Mikulski's second picture of the central parts of the spiral galaxy M100, again taken before and after the servicing mission. Mathematical analysis of the images proved what appeared obvious to the eye: Hubble now fulfilled all the original optical specifications. The new window to the universe had finally been opened.

The Second Servicing Mission

The unqualified success of the first servicing mission led to a continuous stream of press releases about new phenomena and new details of the structure and nature of various astronomical objects. It also led to optimism about the second servicing mission, in which Hubble was to be outfitted with new and better detectors. After all, the scientific instruments used for Hubble had been developed more than 10 years previously and were no longer quite state of the art. Work on the second-generation instruments had already started when Hubble was launched, and the new instruments would contain the technology of the early 1990s.

The two first-generation spectrographs, the Faint Object Spectrograph (FOS) and the Goddard High Resolution Spectrograph (GHRS), would be replaced with two much more modern and complex instruments: the Near Infrared Camera and Multi-Object Spectrometer (NICMOS) and the Space Telescope Imaging Spectrograph (STIS). NICMOS would finally open up the near infrared domain for Hubble, and STIS was a multipurpose spectrograph with an ultraviolet imaging capability. NICMOS can observe the universe in the near infrared with higher sensitivity and better resolution than any other existing or planned telescope. This instrument was developed under the leadership of Rodger Thompson of the University of Arizona. STIS, which can split ultraviolet and visible light into its component colors, features a two-dimensional detector and a thirtyfold increase in spectral information compared to the earlier HST

spectrographs. In addition, it can achieve better spatial resolution by a factor of several hundred. STIS was developed at the Astronomy and Solar Physics Lab of the Goddard Space Flight Center under the leadership of Bruce Woodgate.

On Tuesday, February 11, 1997, at 3:55 A.M. EST, the space shuttle *Discovery* lifted off from Kennedy Space Center at Cape Canaveral, Florida, to begin the second servicing mission for the Hubble Space Telescope. *Discovery*'s crew consisted of seven astronauts: Kenneth Bowersox (commander); Scott Horowitz (pilot); Steven Hawley (arm operator); Mark Lee (payload commander); and Gregory Harbaugh, Steven Smith, and Joseph Tanner (payload specialists). As with Servicing Mission 1, the shuttle reached the telescope in a little less than two days. The final approach demanded a delicate touch from commander and pilot. This time, there were no replacements for the solar panels, and even the slightest bump might have led to irreparable damage to the telescope. After the shuttle had approached within 10 meters of Hubble, Steve Hawley captured it with the manipulator arm and maneuvered it into the payload bay, where it was securely berthed. Upon visual inspection, the space telescope appeared to be in good shape; the solar panels in particular had warped by only a few centimeters. The corrections to the second set, installed during the first servicing mission, had worked.

At 11:34 P.M. EST on February 13, the first space walk began. Mark Lee and Steve Smith removed

the old FOS and GHRS instruments and installed the two new ones, NICMOS and STIS. Before they could begin their activities, however, an unforeseen incident occurred. When the air lock was vented in preparation for the astronauts' exit into the payload bay, the stream of air hit one of HST's solar panels and moved it almost 90 degrees from its rest position. This caused a short-term malfunction of the electronic control system for the solar panels, as well as a considerable scare for astronauts and ground controllers. Luckily, no permanent damage resulted from this incident, and the procedure for venting the air lock was changed for future space walks.

The next day, astronauts Greg Harbaugh and Joe Tanner exchanged one of the Fine Guidance Sensors for the precise positioning of the telescope and

Left: The second visit. The space shuttle *Discovery* is launched on February 11, 1997, to service the Hubble Space Telescope. (NASA)

Astronauts Mark Smith and Steve Lee work on the outer insulating layers of the satellite during the fifth extravehicular activity of the second servicing mission. (NASA)

The Hubble Space Telescope is released into its orbit at the completion of the second servicing mission. (NASA)

installed a new engineering data recorder. During this EVA, the astronauts noticed major defects in Hubble's outer skin, the thermal insulation layer. The extreme temperature changes between day and night, and the constant bombardment with UV radiation from the Sun, had taken their toll. In addition, there were numerous minuscule "craters" from micrometeorites.

On February 16, Lee and Smith performed the third EVA and installed new science data storage components, a new reaction wheel used for pointing the telescope, and a new data interface unit. Because the data interface unit had not been designed for in-orbit maintenance, the astronauts had to disconnect and then reconnect 18 individual cables and connectors, hampered by the somewhat less than sensitive gloves of their space suits. Despite the difficulties, the interface unit was exchanged successfully.

Harbaugh and Tanner exchanged one of the two solar array drive electronic units during the fourth EVA. With all the planned tasks completed, they began to repair the torn and degraded thermal blankets on the outside of the telescope. This insulating skin consists of 15 layers of plastic with an outer layer of aluminized Teflon®. Because the side of the

Astronaut Joe Tanner in front of the Sun and the shadowed Earth. His colleague Greg Harbaugh, who took this picture, can be seen as a reflection in the visor of Tanner's helmet. A task list is visible on Tanner's left arm. (NASA)

telescope that is normally oriented toward the Sun showed several defects, spots, and tears, an additional space walk was planned for the following day. During this five-hour EVA, Lee and Smith used several sheets of packaging material, similar to the thermal blankets, to repair the largest defects, following patterns transmitted from the ground. These efforts were only preliminary measures; based on the astronauts' reports and on images they took of the telescope, a more professional repair is planned for the third servicing mission in 2000.

The total of 33 hours and 11 minutes of extravehicular activities by the payload specialists (again a remarkable astronautic achievement) was only one aspect of the mission. Three times the shuttle's attitude control thrusters were used to lift Discovery and its Hubble payload into an orbit that was 15 kilometers higher than before. The new orbit had an altitude of 620 kilometers at its highest point and 594 kilometers at its lowest; HST now orbited Earth higher than ever before. This higher orbit will be advantageous

The Hubble Space Telescope resumes operation. After the successful completion of the second servicing mission, the satellite is ready for the next three years of astronomical observations. (NASA)

during the coming years, when the expected increase in solar activity leads to an expansion of the Earth's upper atmosphere and consequently to more drag on the huge satellite, resulting in a gradual lowering of its orbit. With all servicing mission activities successfully completed, the Hubble Space Telescope was released on February 19, 1997, to circle Earth another 15,000 or 16,000 times until it is visited again during the third servicing mission.

In the early morning hours of February 21, still in darkness, *Discovery* touched down at the Kennedy Space Center in Cape Canaveral—almost exactly 10 days after its launch. Servicing Mission 2 had not been cheap: the mission had cost a total of $795 million, with the shuttle flight itself costing $448

million and the remaining $347 million allocated to the new scientific instruments, replacement components, and special tools. The annual expenditures for the operation of the Hubble Space Telescope, including the development of new instruments but not the shuttle flight itself, amount to about $230 million. During the 20 years of the space telescope project, from its beginning in 1977 up to and including Servicing Mission 2, a total of $3.8 billion has been spent on the program—a worthwhile expense, as several independent review boards have acknowledged. Given the wealth of new scientific discoveries, they have named Hubble the most cost-effective scientific space project ever.

After the servicing mission, Hubble's regular operations progressed normally at first. Then a higher-than-expected sensitivity of the two new scientific instruments to external radiation became apparent. Whenever Hubble crossed a specific part of the Earth's radiation belt—the South Atlantic Anomaly—during its orbit, the increased radiation background adversely affected several electronic components of the two instruments, leading to bit-flips that in turn caused the instruments to shut down as a safety measure. But a workaround was found quickly: the instruments were simply put into an inactive state while crossing the South Atlantic Anomaly. This measure did not cause any significant loss of observing time, because the high radiation background at the anomaly prevents the detectors from obtaining good-quality data there anyway. With the exception of this problem, STIS and NICMOS appeared to work correctly, until it became obvious that one of the NICMOS cameras could not be focused as expected;

NICMOS and STIS:
Hubble's Second-Generation Instruments

The Near Infrared Camera and Multi-Object Spectrometer (NICMOS) is a combination instrument that makes the near infrared domain of the spectrum available to Hubble for the first time; it is sensitive in the wavelength region from 0.8 to 2.5 microns. This spectral region immediately borders the red side of the visible domain, and it partly overlaps the range of Hubble's WFPC2

in addition, the thermal behavior of the huge dewar containing the cooled cameras did not conform to expectations.

The technique of cooling an instrument in space with frozen nitrogen was fairly new. It now appeared that a "thermal short"—an unplanned mechanical contact between the cold parts of the dewar and the external structure of the instrument—had developed. This caused two things to happen: first, parts of the instrument expanded slightly, moving one of the three NICMOS cameras outside its focusing range; second, and more important, the nitrogen would evaporate much more quickly than expected. Instead of five years of cooled operations from the time of the second servicing mission in early 1997, NICMOS would have only enough nitrogen for less than two years, as calculated from the actual venting rate. The expansion of the dewar subsided in March 1997 and even reversed somewhat, but not enough to

camera. NICMOS consists of three independent cameras with different pixel sizes (0.043, 0.075, and 0.2 arc second) and thus with different fields of view and angular resolution.

Each camera features a mercury-cadmium-telluride array of 256 by 256 pixels and has its own filter wheel. The filter wheels carry not only color filters but also polarizers, coronographs, and grating prisms (or "grisms"). NICMOS is Hubble's first cryogenically cooled instrument; according to its specifications, it was supposed to be cooled to 58 K by frozen nitrogen for five years. Unfortunately, this goal was not realized. The total cost for the development of NICMOS amounted to $105 million.

NICMOS is particularly suitable for the observation of objects in the early universe, of stars in and behind dust clouds in our galaxy, and of cool objects such as brown dwarfs. One target of NICMOS (and of STIS as well) will be the Hubble Deep Field, both the original northern field and the planned southern field. In principle, NICMOS could observe protogalaxies with a redshift of 14(!)—if they indeed exist.

The Space Telescope Imaging Spectrograph (STIS) is a very complex, broad-band, multi-purpose spectrograph with 13 different spectrographic and 8 other supporting operating modes. The development of STIS required $125 million. It has two detectors that combine to cover a spectral range from 115 nanometers to 1 micron, or from the ultraviolet through the visible range to the beginning infrared. Two so-called MAMA detectors (short for Multi-Anode Multi-Anode Channel Arrays) are used in the ultraviolet from 115 to 310 nanometers. Each detector consists of 1024 by 1024 minute image intensifier tubes. Each UV photon knocks electrons from the photocathode; the photons are multiplied by a factor of 400,000 in their descent through the micro-

pores, induced by a potential difference between the top and bottom. In this way, the spatial coordinates and time of arrival of each photon in the array can be registered electronically. The STIS does not even require a mechanical shutter.

For longer wavelengths (from 305 nanometers to 1 micron), a traditional charge-coupled device (CCD) chip is used. The chip is similar to those in Hubble's WFPC2 or ground-based astronomical cameras and is a distant cousin of those used in modern camcorders and digital cameras. A filter wheel with 65 positions contains entrance apertures and slits, as well as filter combinations for spectroscopy and calibration. The main advantage of STIS over Hubble's old spectrographs lies in its two-dimensional detectors. Instead of dispersing the light from a single small aperture into a one-dimensional spectrum, the light from an entire long slit is spread out into a two-dimensional "image" (with one spatial coordinate parallel to the slit and one wavelength coordinate). This arrangement allows much higher observing efficiency and makes it possible to obtain spectral information about extended objects—such as galaxies, nebulae, and planets—in short times. An important task will be the investigation of intergalactic absorption lines in the spectrum of quasars. A coronograph to observe faint structures around bright stars is also available.

bring the one camera into the same focusing range as the other two. That camera can still be used for specific campaigns, however, by refocusing the entire telescope (unfortunately, though, this changes the focus of all the other instruments away from its optimum value).

To make the best use of the available time, the entire observing program was revised. Much more time than originally planned would be used to observe with NICMOS, and many of the programs using the other instruments would be postponed until after NICMOS runs out of nitrogen. Since mid-1997, almost half of HST's observing time has been dedicated to this instrument, and almost all of the planned NICMOS research programs will be completed by late 1998. But scientists and engineers are already planning to try to resurrect NICMOS after it has run out of cryogen. It might be possible to install an external, active cooling system for NICMOS during the third servicing mission in the spring of 2000, thus extending its lifetime until the end of the HST mission. To keep the block of nitrogen frozen before the launch of the second servicing mission, NICMOS has a system of pipes, through which cold liquid helium was pumped. These pipes are still in place, although they are not currently being used. But they could be used to pump cryogen through NICMOS and cool it back down again. Current plans foresee the use of neon as a cooling agent. This would allow the

NICMOS detectors to be cooled to 70 K, close to the temperature of 58 K achieved with the frozen nitrogen. A final decision on the installation of the NICMOS cooling system during Servicing Mission 3 has not been made at the time of this writing.

The adjustment and calibration of NICMOS and STIS had not yet been completed when the first test images were made public. Two "Early Release Observations," one from the new infrared camera and one from the imaging spectrograph, were released on May 12, 1997. These images, however, did not receive much media attention. Major publicity remains the domain of WFPC2, with its wonderfully sharp and detailed images that make it to the front pages of newspapers and into network news. But the scientific community was quite impressed by these early observations, and in January 1998, the first results based on NICMOS and STIS data appeared in the astronomical literature.

Let us now follow Hubble on its observations of the universe, as we did in our first book, which described the results from the years 1990 to 1995. We will begin at the edge of the universe with questions about its age and size, continue with distant galaxies, visit the stars and nebulae of our Milky Way, and arrive back in our own solar system. Details of the next planned servicing missions, the future of Hubble, and its successor can be found in the last part of this book.

Part 2

To the Edge of the Universe

The Basic Questions of Cosmology

The universe is more than just a collection of cosmic objects: it has geometric characteristics, and exploring them is one of the main tasks of cosmology. For instance, scientists want to know the average density of matter in the universe—that is, how much mass per volume exists if all planets, stars, gas clouds, and so on, were "ground up" and equally distributed in space. We can follow this thought experiment easily and can even realize that the density of matter determines the fate of the entire cosmos: if the matter density is too low, then the universe will expand forever.

But our imagination cannot easily follow all the consequences of the cosmic density. It determines, for example, whether space as a whole is curved and, if so, how. Today, the universe expands—but its density; curvature; and an even less comprehensible value, the cosmological constant, determine how it will behave in the future. Will the expansion eventually come to an end, or even reverse itself? Such questions have been unanswerable for a long time, but that has begun to change in the last years of the twentieth century. Sophisticated observing techniques make it increasingly possible to *measure* the fundamental numbers of the universe. The Hubble Space Telescope and large telescopes on the ground have made cosmology an observational science.

If we want to understand the structure of space, we must explore its content—and luckily, the content appears to be fairly well organized. Matter has distributed itself, by whatever means, into a hierarchical structure. The basic units are galaxies and stellar systems, in which billions of stars have congregated into spheres, disks, or irregular shapes, bound together by their gravitational pull. Our own star, the Sun, is part of such a galaxy, the Milky Way. And galaxies themselves form larger units, galaxy clusters and superclusters, that are embedded in gigantic bubble-shaped structures. The galaxies and the clusters they form are the walls of these bubbles, surrounding virtually empty interiors. This was one of the fundamental astronomical discoveries of the 1980s. It was not easy to arrive at this conclusion, for we are part of the cosmic whole. From a single undistinguished point, we look around and attempt to fathom the grand plan of the universe.

Our view into space includes close suns as well as distant galaxies. A look into the depths of space is at the same time a look into the past of the universe, for the speed of light is not infinite. By cosmic standards, in fact, light's speed of 299,792.5 kilometers per second can be considered rather slow. We read the universe as a book with unnumbered pages that have been totally scrambled. It is the job of the cosmologist to arrange all the pages into their proper sequence. The scope of this task makes cosmology both fascinating and difficult. The determination of distances is one of the most fundamental tasks in astronomy, but direct methods that yield an absolute number for the distance of a star are available only within a very small area of our own galaxy. All other distance determinations are relative, describing only how much farther away one celestial object is from us than another. Only the combination of many such measurements can provide the absolute distance to a faraway star or distant galaxy, whose light has traveled for millions or even billions of years to reach us.

The universe contains a veritable zoo of galaxies. Some galaxies are large and some are small. Some galaxies continually produce new stars from dust and gas; others consist only of old stars, their gas having been swept away. Galaxies can be classified based on their appearance: there are spiral, barred spiral, elliptical, and irregular galaxies. But this classification is only a first step; astronomers, after all, are not just collectors, content to note the appearance of various galaxies: they want to know more about their structure and evolution. The astronomer Edwin Hubble, for whom the space telescope is named, was among the first to propose a classification of galaxies, and his scheme is still more or less in use today.

Briefly, Hubble's classification scheme is as follows. *Spiral galaxies* are characterized by a bulge in the middle, surrounded by stars and interstellar clouds that appear as spiral arms. These galaxies are further subdivided based on how tightly the arms are wound. A *barred spiral galaxy* is a spiral galaxy with a bar of stars crossing through the central bulge. *Elliptical galaxies* have no spiral arms and are more or less elliptically shaped. They are subdivided based on how round they look. Finally, some galaxies cannot be classified as spirals, barred spirals, or ellipticals, and these are called *irregular galaxies.*

Hubble was also the first astronomer to begin to comprehend the true size of the universe. At the beginning of the twentieth century, the size of our own galaxy was understood to be several tens of thousands of light years, but it was not known whether this represented a major part of the cosmos or only a tiny dust speck in the ocean of the universe. Faint spherical or spiral objects could certainly be seen in the largest telescope of the time, and these objects *could* be distant stellar systems. But they could also be much smaller objects within our own Milky Way.

In 1923, Hubble achieved a major breakthrough when he discovered pulsating stars, which also exist in our own galaxy, in one of the possibly distant stellar systems, the Andromeda Nebula. A fascinating law had been discovered about these special stars, called Cepheids: the period of their pulses is directly proportional to their true or absolute brightness. By observing the pulsational period of a Cepheid, one can calculate its absolute brightness—and once the brightness at which the star appears *to us* has been measured, its *distance* can be derived without further problem! We had discovered a "standard candle" that could be used to determine distances in space directly, without intermediate steps. The Cepheids Hubble discovered showed him that the Andromeda Nebula is so far away that it must be a separate galaxy. The universe had suddenly become much larger.

In 1929, after Hubble had determined the distance to a large number of galaxies, he discovered a linear relationship between the distance and the redshift in the spectra of these galaxies. This relationship, called the Hubble law, states that the more distant a galaxy is, the faster it recedes from us. The redshift of galaxies is therefore an enormously useful tool in bringing order into our book of the universe: The farther away a galaxy is from us and the longer its light has traveled to reach us, the more this light has been altered by the expansion of the universe — it has been shifted to longer, redder, wavelengths.

We can measure this cosmological redshift for a galaxy by exploring its spectrum. Well-known spectral lines caused by various chemical elements (such as hydrogen, calcium, and iron) do not appear at the wavelength at which we measure them in a terrestrial laboratory. Instead, they are shifted toward the red part of the spectrum. A large redshift, therefore, indicates a page near the beginning of our cosmological book. And the more we go back in the book, trying to decipher its early pages, the more mysterious the universe appears.

Unfortunately, it is not possible to measure redshifts for all galaxies. Such a task would take in-ordinate amounts of observing time on the largest telescopes, and astronomers count themselves lucky to obtain a few nights of observing time per year. Even large-scale projects begun in 1997 at several observatories where large telescopes are dedicated solely to galaxy spectroscopy will be able to determine exact redshifts for only a fraction of these distant islands in space. It will not be possible to reconstruct the complete history book of the universe. But there are a number of tricks that can be used to get at least a rough estimate of the distance of a galaxy.

The farther we look into space, the fainter the galaxies appear. Not only are they more distant, but their blue light, usually fainter than their light in the visible range, is now being shifted into the visible range as a result of the cosmological redshift. Galaxies normally do not radiate at all in the far ultraviolet range. As we observe galaxies with high redshift using HST's color filters, we sometimes notice that they are no longer visible in the bluest filters. Such "dropouts" indicate galaxies with very high redshift — extreme cases even exhibit a dropout in red light. In addition to spectroscopy, there are other, more convenient (if not as accurate) methods to determine redshifts and corresponding distances.

A View into the Depths of Space and Time: The Hubble Deep Field

What would happen if the Hubble Space Telescope were pointed at the same spot in the sky for a few weeks to obtain an extremely deep image of the cosmos? No team of astronomers would get this amount of observing time from the group of independent reviewers who select the proposals for execution—there are simply too many interesting proposals for the limited observing time available. But on Hubble as well as in some ground-based observatories, a fraction of the observing time is made available to the director to distribute at his or her discretion; usually, this time is used when unforeseen events occur or the opportunity for unique observations arises. At the end of 1995, Hubble's discretionary time was used to observe one small area in the sky for 10 days. The main characteristic of the target area was that it contained "nothing"—no star, quasar, or conspicuous radio or infrared source.

The target area, about one-thirtieth the size of the full Moon, is located in the constellation Ursa Major, also known as the Big Dipper, a small distance north of the star Delta Ursae Majoris (the star at which the handle is attached to the cup of the Big Dipper). Known as the Hubble Deep Field (HDF), this tiny part of the sky has become an astronomical celebrity. During 10 contiguous days between December 18 and 28, 1995, a total of 342 exposures in four colors were taken with the WFPC2. Over the next few weeks, an international team of astronomers processed and calibrated the images, added the individual color exposures together, and finally combined them into a single color image. On January 15, 1996, it was proudly presented to the astronomical community and the general public.

It was not only the specialists who were impressed. The HDF showed objects—almost all of them distant galaxies—that were 3 to 15 times fainter than had been observed by astronomers in any other deep field up to that time. In the red region, the 30th astronomical magnitude had been reached for the first time—a number that would excite any astronomer. The *larger* the magnitude, the *fainter* the star or galaxy appears in the sky. And a 30th magnitude galaxy appears in the sky two-and-a-half billion times fainter than the faintest star visible to the naked eye on a very dark night. The Hubble Space Telescope had opened another "new window on the universe." But what does this deep view into the depths of space show?

What would happen if the universe were eternal, were of infinite extent, and everywhere contained luminous matter (stars and galaxies)? This question was posed very early, and the German physician and hobby astronomer Wilhelm Olbers gave a fitting answer in 1823: the night sky would have to be extremely bright, because wherever a terrestrial observer looked, his view would hit the surface of a radiating star at every point in the sky. Because this strongly contradicts our experience that the night sky is dark, this question is called Olbers' paradox. Olbers himself had proposed a first (but erroneous) explana-

Thousands of galaxies and a handful of stars in the Hubble Deep Field. The faintest objects in this image are 4 billion times less bright than the weakest stars visible by eye on a dark night. This image is a composite of 276 of 342 exposures taken in the blue, red, and infrared. Only one of the four parts of the image is shown here. (Williams and NASA)

How "Deep" is the Hubble Deep Field?

How deep into space does the Hubble Deep Field—that record-setting view at one and only one location in the sky—reach? This question cannot be answered easily, because redshifts could be obtained directly by spectrographs only for the brightest of the more than 1500 galaxies in the image. Because the individual exposures were made in four different colors, however, the redshift of the much fainter galaxies can be estimated. Characteristic details in the spectra of galaxies move toward longer wavelengths with increasing redshift. In this context, the so-called Lyman limit is important. Neutral hydrogen in intergalactic space radically suppresses the light from distant galaxies below a certain wavelength in the ultraviolet, forming a characteristic cutoff—the Lyman limit—in the spectra of the galaxies. In addition, the remainder of the UV spectrum is dampened by a veritable forest of absorption lines from hydrogen in intergalactic clouds. Both effects apply to distant galaxies regardless of their individual characteristics, because only the space between them and us is responsible for these effects.

These effects make it possible to attempt the derivation of redshifts in the Hubble Deep Field from CCD images in various filters. This technique worked fairly well in computer simulations; for the majority of them, the technique of using different color filters yielded practically the same redshift that had been used at the beginning of the simulation. Within only a few years, the technique has become the standard for determining or estimating the distance of a large number of faint galaxies in a short period of time.

The technique has been applied to the Hubble Deep Field. Redshifts were obtained for a total of 1104 objects: 367 galaxies showed a redshift between 0 and 1, 512 a redshift between 1 and 2, 135 a redshift between 2 and 3, 30 a redshift between 4 and 5, and 2 a redshift between 5 and 6; four galaxies had redshifts even larger than 6! With redshifts of up to 2.5, the spectra are very similar to those of nearby galaxies, but shifted according to their velocity. Between redshifts of 2.5 and 4, however, the objects become bright at wavelengths of 814 and 606 nanometers and are observable at 450 nanometers, but are too faint at 300 nanometers: Here, the intergalactic hydrogen begins to block the light.

Above a redshift of 4, the galaxies become unobservable at 450 nanometers, and above a redshift of 6, they cannot be seen even at 606 nanometers; they can be seen only in the near infrared. The multicolor technique cannot deliver

The third dimension of the Hubble Deep Field is the depth of space. All numbers in the large image, which shows the entire HDF, are redshifts of galaxies, as measured with the 10-meter Keck I telescope. The faintest galaxies still evade spectroscopic observation, but their distances can be estimated from their colors. The object marked in the detailed view is visible only in the reddest color filter, indicating a galaxy with extremely high redshift. (Lanzetta & Yahil, Williams, Keck Observatory, and NASA)

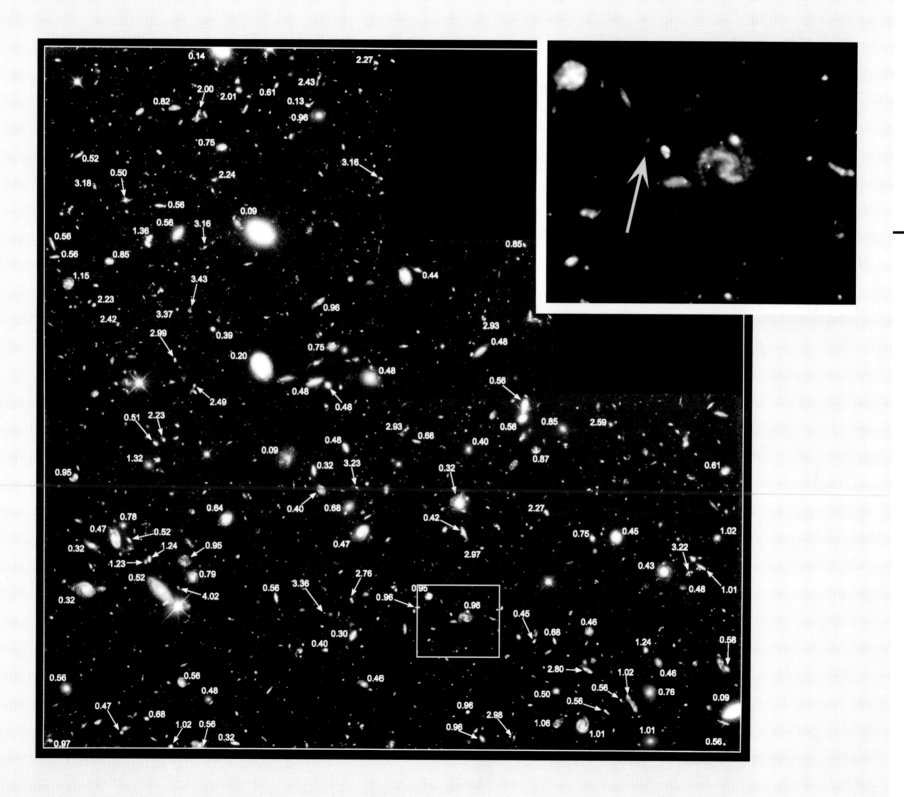

irrefutable proof that these points of infrared light are indeed very distant galaxies, but this is still the most probable explanation. In addition, some of these objects appear to be spatially resolved, or at least not starlike. Assuming that their high redshift values are correct, the luminosity of these galaxies corresponds to 10^9 to 10^{10} solar luminosities. Such luminosities would be moderate when compared to luminous galaxies, but they are comparable to starburst galaxies in our vicinity, which undergo a phase of strong star formation. In these high-redshift galaxies, we may be witnessing the first generation of star formation in the universe, and these points of infrared light are parts of larger galaxies: Our own Milky Way may have started its development in this way.

The only question is when? At such large redshifts, such statements as "x percent of the age of the universe" or "y million years after the Big Bang" depend heavily on the preferred model of the universe! In the mathematically simplest case (with a Hubble constant of 100 km/s/Mpc, critical density, and no cosmological constant), we would look at these galaxies over more than 95 percent of the time since the Big Bang and observe processes that occurred less than a billion year after it. Such a view of the universe, however, becomes less and less likely, based on recent results. In a nonstandard cosmological model with a positive cosmological constant, which has become more probable over the last years, several billion years may have elapsed between the Big Bang and the time at which we are seeing these high-redshift galaxies with $z = 6$. The existence of galaxies with such high redshifts would be an argument in favor of this cosmological model, because more time would have been available for their formation and development.

tion: dust between the stars should shield us from the radiation of more distant objects.

Although this paradox did not spark much interest among professional astronomers at first, the author and journalist Edgar Allan Poe attempted a solution in his cosmological sketch "Eureka": the universe is finite in time, so that the light from distant stars has not had enough time to reach us. And he was right! But it was not until the late 1980s that cosmologists could be certain about this fact. It is primarily the finite age of the universe, not its expansion, that causes the night sky to be dark. The expansion of the universe, which shifts the light of distant galaxies toward the deep red, could only

The wealth of shapes of faint galaxies in the Hubble Deep Field is demonstrated by three excerpts. In addition to classic elliptical and spiral galaxies, several exotic and irregular shapes and colors are visible. (Williams and NASA)

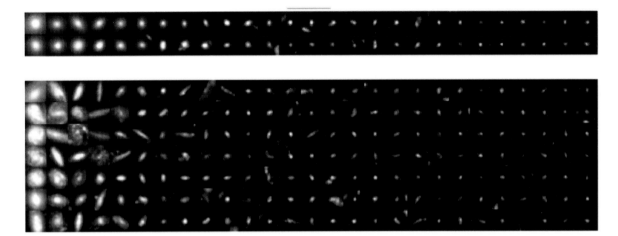

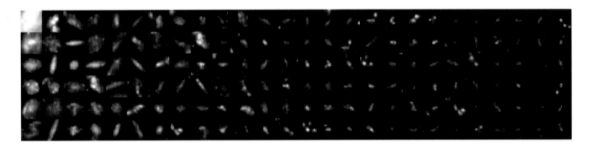

The galaxies of the Hubble Deep Field, sorted by their morphological type, or appearance. Their brightness decreases from left to right. (Driver)

Reading the Book of the Universe

The rapid release of the original data from the Hubble Deep Field to scientists all over the world has been a very fruitful decision. Since then, a steady stream of new research about the deepest view into space has been published. The impression from simple optical inspection of the HDF image, that more irregular galaxies must have existed in the past, was confirmed within a few weeks using statistical techniques. Thirty to forty percent of the distant galaxies appear to be unusual or deformed, compared to only a few percent in today's cosmos. Additionally, the trend toward more irregular galaxies, which had already been seen in Hubble images obtained with shorter exposures in 1995, increases significantly for fainter and more distant galaxies. The

early universe, which has been made accessible by the Hubble Deep Field, is quite different from today's universe, so that Edwin Hubble's venerable scheme of galaxy classification is no longer applicable.

Even the simple counting of galaxies provided valuable insights. The number of galaxies increases dramatically the deeper we look — that is, the fainter the objects we observe. Even allowing for the development of galaxies over time and for distant, young galaxies to be brighter on average than today's objects, we are still left with more than four times as many galaxies as would be predicted for a universe at critical density. This result could be an indication of an open universe with less than critical density. However, some of the reports in the newspapers shortly after the presentation of the Hubble Deep Field,

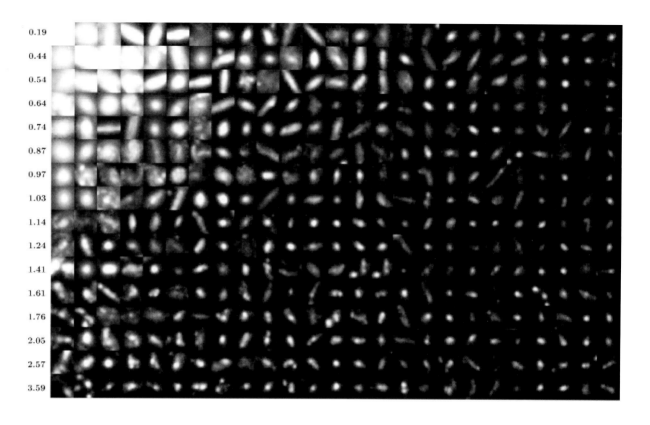

0.19
0.44
0.54
0.64
0.74
0.87
0.97
1.03
1.14
1.24
1.41
1.61
1.76
2.05
2.57
3.59

The galaxies of the Hubble Deep Field, sorted by redshift (increasing from top to bottom) and brightness (decreasing to the right). The evolution of galaxies can now be seen in general terms, but the cosmological details require extensive computational analysis. (Driver)

which said that the image had shown the universe to contain not 10 billion but 50 billion galaxies, were wrong: this number had already been derived from earlier HST observations in 1993. The Hubble Deep Field can even provide information about the number of faint red stars in our own galaxy. Galaxies and stars up to 26th magnitude can be easily distinguished—and no faint red stars can be found. Therefore, such stars do not provide an important contribution to the dark halo of our galaxy.

None of these analyses require the exact distance of individual galaxies (the true third dimension is still missing). But using the sophisticated analysis of the brightnesses of galaxies in different colors, their redshifts can be estimated rather well. And redshifts for individual galaxies can be measured directly using the largest telescopes on Earth. Again, a surprising result occurred: 40 percent of the distances determined for 140 galaxies using the 10-meter Keck telescope are concentrated around six distinct values! This result indicates that galaxies throughout the universe are concentrated in bubble- or sheetlike structures with great voids between them. We must realize, however, that the Hubble Deep Field with its tiny field of view is probing the universe only in a single direction. In principle, the galaxies could be located in isolated clumps, rather than within extended sheets. But modern models of the formation of structure in the universe are predicting such immense, thin sheets or bubbles on the grandest scales.

account for making the sky darker by a factor of two, at best—it would still be as bright as daylight. The fact that the night sky is dark finally proves that the universe began to exist a finite time ago. If only all other cosmological problems were that obvious!

Hubble's view had penetrated deeper into space than any previous attempt by astronomers—and depth corresponds to fainter and fainter celestial objects. Only a few stars of our own galaxy are visible in the Hubble Deep Field image; their number is far exceeded by the number of distant galaxies. The brightest object in the field is a star with diffraction spikes of about 20th magnitude, and the faintest visible objects—of almost 30th magnitude—are ten thousand times fainter. The HDF may not be the grand book of knowledge from which we can read the history of the universe, but it can help to confirm the truth of the "Tales of the Universe" that theoreticians have devised and allow us to revise such theories.

But have we seen most of the luminous matter in the Hubble Deep Field, or is a significant number of galaxies hiding among the visible systems in the electronic noise, too faint to be seen by Hubble's WFPC2? We can answer this question by using a mathematical method called autocorrelation. We can analyze whether the brightness of individual pixels between the galaxies of the Hubble Deep Field is randomly distributed or even fainter galaxies are lurking there, just not recognizable by the human eye. This analysis was completed in 1998, and the results are unambiguous. In the Hubble Deep Field, we indeed see essentially the entire luminous universe in the optical wavelength range in that direction! There may be more galaxies hidden among the 2000 recognizable ones, but their combined visible light would amount to only a small fraction of the light of the observed systems.

We can*not* see the entire cosmos with Hubble, however long an exposure we may take, since significant contributions appear in entirely different wavelength regions. We learned this from observations made with another astronomy satellite that were subjected to even more complex data analysis.

In the early 1990s, NASA's Cosmic Background Explorer (COBE) had scanned the entire sky many times with telescopes that were sensitive to the far infrared, the transition region between light and radio radiation. They saw primarily warm dust, heated by stars—stars that this dust hides in the visible range. Immediately visible in these survey charts of COBE's Diffuse Infrared Background Experiment (DIRBE) were dust components in our own solar system and in the Milky Way. It took seven years to understand and isolate all these dust components and to subtract them carefully from the remaining images. By the beginning of 1998, several independent teams of astronomers were certain: even after subtracting all the known dust, a "background glow" remained at a wavelength of 140 microns (0.14 mm). It is still unclear where in the universe the dust that causes this infrared glow is located, but the brightness of this totally uniform glow from all directions

A detail of the Hubble Deep Field in false color, about 40 arc seconds on a side. Individual galaxies and stars appear in red, but the focus of this image was the "empty" space between them. Statistical analysis shows that this space contains only noise, not more galaxies. Hubble has found most of the galaxies that exist in this area. (Vogeley and NASA)

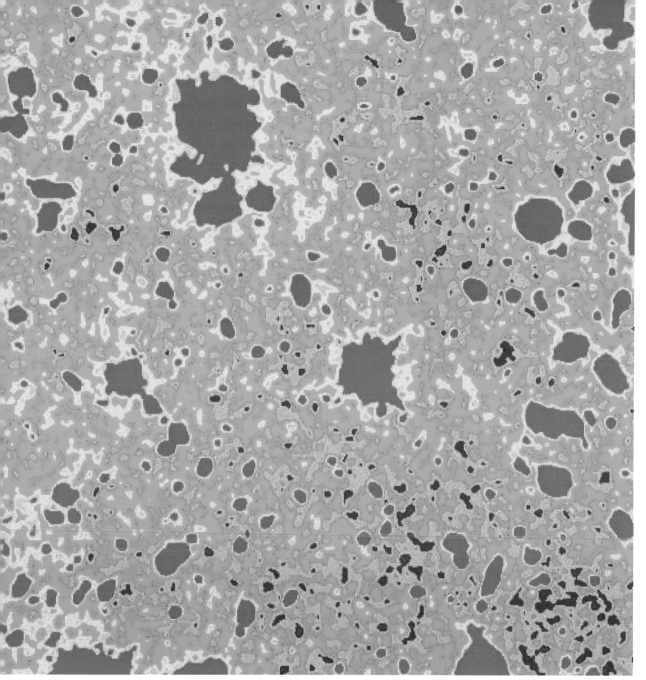

reveals much information about the true history of star formation in the universe.

About two-thirds of the light from all stars is hidden from us, and Hubble as well, because it is immediately absorbed by dust. Now COBE has found this light, and the history of the universe can be written much more completely. The history of the formation of new stars in the course of cosmic evolution, for instance, must be rethought. We cannot rely solely on observations of galaxies in visible light—all wavelength ranges must be taken into consideration! We will understand the Hubble Deep Field fully only after it has been observed in all available wavelength ranges. Because the Hubble Deep Field image has stimulated astronomy more than any other single image in its history, an additional field—the Hubble Deep Field South (HDF-S)—will be observed with WFPC2 and with the new NICMOS and STIS instruments in October 1998. HDF-S is located in the constellation Tucan, not far from its brightest star, Alpha Tucanae.

The Search for Cosmic Numbers

The past and future of the universe can be described surprisingly accurately, once only a handful of numbers have been determined. This is one of the unexpected results of the theoretical investigations in cosmology at the beginning of the twentieth century. Some of these fundamental equations are as simple as can be. The distance and the redshift of a galaxy (due to its cosmic expansion velocity) are connected by the simple relationship $v = H_0 d$, where v is the velocity in kilometers per second, d is the distance in megaparsecs (1 megaparsec = 3,260,000 light years = 3.1×10^{19} kilometers), and H_0 is the Hubble constant, the measure for the rate at which the universe expands. The determination of this number is theoretically very simple. The redshift is plotted against the distance for a number of galaxies; this plot will show a linear relationship (the Hubble law), and the slope of that line gives the value for the Hubble constant.

The accuracy of the value of the Hubble constant depends on the precision of the redshift measurement (which is usually quite high) and on the precision of the distance measurement (which is usually the problem). In 1929, when Edwin Hubble plotted his "Hubble diagram" for the first time, he derived a value of 530 kilometers per second per megaparsec for his constant. That turned out to be too large by at least a factor of six. Today's determinations of the Hubble constant almost always lie between 50 and 100, and often between 60 and 80, kilometers per second per megaparsec. If we assume a value of 100 for the moment, Hubble's law indicates that a galaxy at a distance of 1 megaparsec moves away from us at a velocity of 100 kilometers per second, a galaxy at a distance of 2 megaparsecs at a velocity of 200 kilometers per second, and one at a distance of 10 megaparsecs at a velocity of 1000 kilometers per second.

Thus, the inverse of the Hubble constant must be a measure of the age of the universe. If we could let time run backward, all galaxies would move toward, rather than away, from us at their present speed. At a certain point in time, they would all collide, and that point would correspond to the Big Bang—the birth of the universe. But the situation is not quite that simple. In addition to the Hubble constant, there are other quantities that describe the structure and evolution of the universe: the acceleration parameter q, the density parameter ω (omega), the pressure p, the cosmological constant λ (lambda), and the curvature of space; all these quantities are connected by mathematically rather complicated relations. None of them can be derived from first principles; they all must be measured—and nature does not reveal them easily.

The acceleration parameter describes whether and how much the expansion of the universe has slowed down since the Big Bang. The density parameter indicates whether the density of the universe is high enough that the gravitational forces between its parts can bring the expansion to a halt at some time, leading to a later collapse. The critical density is the density that would bring this expansion to a halt after an infinitely long time. The pressure is composed of radiation pressure and the movement of galaxies; in the present, this quantity can be neglected. The cosmological constant indicates the existence of a repelling force increasing with distance between two objects, a similar attracting force (in addition to gravitation), or none at all. If there is no repelling or attracting force, the cosmological constant is zero. The curvature of space can be described by the values −1, 0, or +1: if the curvature is 0, space is flat and Euclidean; if it is +1, space is spherically, or positively, curved, like a sphere; and if it is −1, space is hyperbolically, or negatively, curved, like a saddle.

One of today's widely discussed theories, the theory of the inflationary universe, expands the universally accepted standard model of the universe. In its simplest form, the theory states that today no additional repelling force exists (the cosmological constant is zero), but that the very early universe had been "blown up" to several billion times its original diameter by the existence of a cosmological constant. It is postulated that the universe today is flat, that the theorems of Euclidean geometry are fully valid, and that the universe has a density equal to the critical density. But the observable (luminous) matter in the universe contributes less than 1 percent to the critical density, and nonluminous matter (very cold or very hot thin gas, brown dwarfs, and black holes) contributes only up to 10 percent. The remaining 90 percent would have to consist of dark matter—possibly very massive elementary particles that interact very little with the "normal" matter of the universe. All this is conjecture, however, and it could well be that future generations will laugh about this idea of a flat universe, just as we laugh today about the medieval idea of a flat Earth. Indeed, a different kind of universe appears to be indicated by new observations, including those from the Hubble Space Telescope.

The Continuing Search for the Hubble Constant

Independent of the applied cosmological model, the equations for the age of the universe yield values between $(2/3)(1/H_0)$ and $1/H_0$. The first, smaller value results when the density of the universe is equal to the critical density, the second when the density is much lower. For $H_0 = 100$, the age of the universe would be between 7 and 10 billion years; for $H_0 = 50$, the age would be between 14 and 20 billion years. Should the cosmological constant be positive and large, however, then the Hubble constant would vary considerably over the history of the universe, being small for an extended period of time and then increasing to its present value. Such a universe could then be much older than indicated by the inverse of today's Hubble constant. In that case, a universe with a current value of $H_0 = 80$ could still have an age of 20 or 30 billion years, or even more. Discussions about the "best" cosmological model

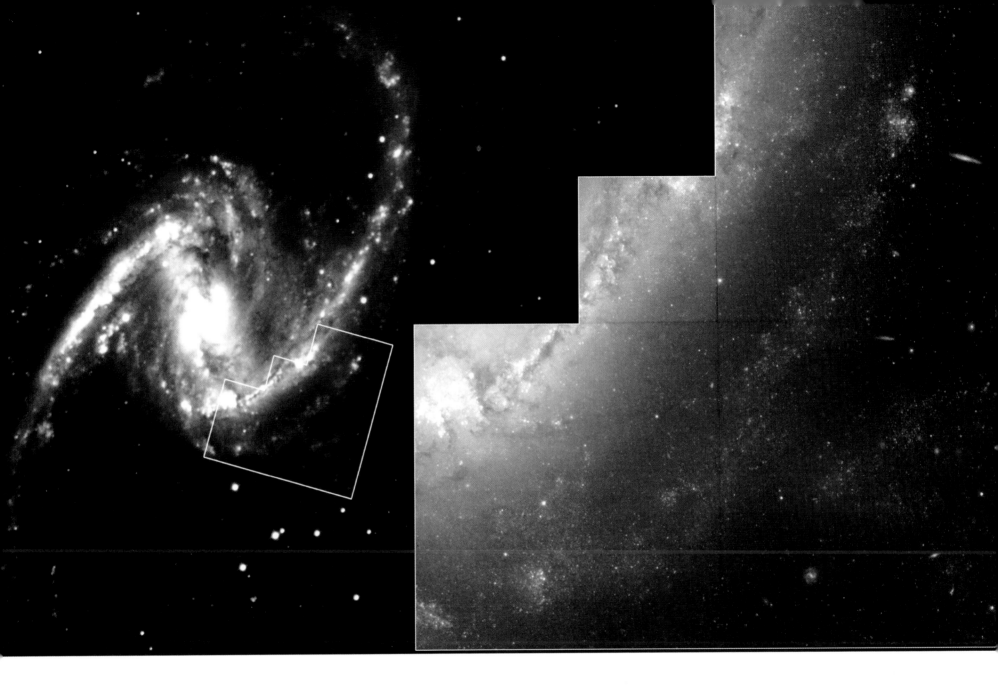

The galaxy NGC 4639 in the Virgo cluster. A supernova appeared in this galaxy in 1990, and Hubble was able to observe a number of Cepheids as well. Several of the blue dots in the outer regions of the galaxy belong to this class of variable stars. Their presence made this galaxy an additional step in the cosmological distance scale. (Sandage et al. and NASA)

Successful Cepheid observations in the Fornax cluster of galaxies. In this barred spiral galaxy, NGC 1365, the H_0 Key Project discovered about 50 Cepheids, which allowed astronomers to determine the distance of the galaxy and therefore also of the entire cluster. The relative motions of the cluster's galaxies can be determined more easily than in the Virgo cluster, leading to a better derivation of the Hubble constant. (H_0 Key Project and NASA) could go on forever—until we begin to measure the critical quantities, such as the Hubble constant, the density, and the cosmological constant. In this area, the Hubble Space Telescope has been able to advance astronomy more than any other instrument.

The determination of the Hubble constant revolves around the problem of knowing the absolute magnitude of a distance indicator exactly and around the need to probe far into space to measure the true Hubble flow of the galaxies—in other words, the expansion of the universe, manifested in the cosmological redshift. Unfortunately, there are superimposed individual velocities for galaxies, which cause additional Doppler shifts and introduce scatter into the data for determining the Hubble constant. Moreover, the requirements for well-known distance indicators and for great depth contradict each other! Reliable distance indicators that can be calibrated in the vicinity of our Sun are not bright enough to be found in distant galaxies with redshifts, which are caused mainly by the Hubble flow. Even in the age of the Hubble Space Telescope, there is no way around an unfortunate compromise: using "close" distance indicators, we must determine the distance to nearby galaxies, where we can find different, much brighter distance indicators. And these are then applied to more distant galaxies, which are part of the Hubble flow. Now we have reached our goal and have built something that is commonly called a cosmological distance scale.

Today, we know that Edwin Hubble's distance determinations in 1929 contained large errors. We also know today how to do it better—but as we watch the various groups of scientists debate about distances and the value of the Hubble constant, it is clear that we do not know how to do it right. Do we rely on good, bright distance indicators, the supernovae (exploding stars), and calibrate their brightness using the well-known Cepheids (pulsating stars)? Do we use the Cepheids to calibrate an entire arsenal of bright distance indicators to determine the distance to faraway galaxies? Or do we replace the Cepheids with other close-by distance indicators, which are equally precise or even better? One of HST's main tasks has been the observation of Cepheids in other galaxies to determine their distance with greater accuracy. If the period of a Cepheid is known, then its true absolute brightness can be derived, thanks to the period-luminosity relation exhibited by this class of stars. And if this calculated absolute brightness is compared to the observed apparent magnitude, the distance of the star can be determined—and therefore also the distance of the galaxy in which the star is located.

Aside from Cepheids, supernovae of type Ia are used as standard candles for jumps to greater distances. Such supernovae do not occur very often in the vicinity of our Milky Way, however, and so we have to calibrate the supernova distance scale with that found from Cepheids. Thus, one method overlaps with the other, and depending on the chosen indicators and the weights assigned to them, different groups of scientists find different values for the Hubble constant. Recently, the various opposing camps have at least closed the gap somewhat, so that most of the determinations yield a Hubble constant between 55 and 85 kilometers per second per megaparsec, corresponding to an age for the universe of between 7 and 17 billion years—unless the cosmological constant is large, in which case the values would be larger. The hope that observations with the Hubble Space Telescope would end this controversy within a few years has not been fulfilled. One of HST's key projects uses considerable amounts of observing time to measure Cepheids in many nearby galaxies and attempts to build a distance scale from many interlocking components. According to some members of this project's large team, the "final" value for the Hubble constant will most likely be slightly below 70 kilometers per second per megaparsec.

But other astronomers who rely solely on supernova explosions as standard candles for larger distances are convinced that the "true" value will be below 60. And yet another faction maintains that the Hubble constant is larger than 80. In principle, this situation is no different from the "50 versus 100" controversy of 20 years ago. But there is one practical difference: all the groups begin the determination of their distance scales almost exclusively with Hubble Space Telescope observations of nearby galaxies, whether of pulsating Cepheids or of certain types of bright giant stars. They still arrive at different end re-

sults—at least for now. But the discrepancies have decreased: 60 is now favored instead of 50, and 73 instead of 100 by the other side.

The Future of the Universe: Perpetual Expansion

Astrophysics still does not have the absolute measure of the universe, but that does not mean that clear statements about fundamental characteristics of the universe are impossible. The Hubble Space Telescope plays an important role in the determination of the density of space, and therefore the acceleration parameter, and the cosmological constant. The most decisive method for measuring the geometry of space, currently advanced by at least two teams of scientists, is based on supernovae of type Ia. These stellar explosions are not all of the same brightness, but the observed development of the brightness after the explosion allows for corrections, so that the *relative* distances between these stars and their parent galaxies can be determined with great accuracy. Even if the absolute distances remain controversial, the relative numbers are sufficient to determine whether or not the expansion rate of the universe (that is, the Hubble constant) has changed over the course of billions of years—and if so, in which direction.

For these investigations, the corrected greatest brightness reached by the distant supernovae is plotted against the redshift of their parent galaxies. The shape of the resulting curve is determined by the density of the universe and by the cosmological constant. The differences between these curves become significant only for redshifts larger than 0.4; therefore, the discovery in 1997 of a supernova in a galaxy with a redshift of 0.83 was cause for celebration. This additional data point improved the accuracy of the measurements considerably, and it became increasingly likely that the density of the universe was much smaller than the critical density. But this was only the beginning: after long preparations, groups of scientists are now able to observe supernovae in distant galaxies much more frequently, and Hubble programs are in place that allow the observation of such supernovae on short notice. Two nights every month, the 4-meter telescope of the Cerro Tololo Interamerican Observatory in Chile is dedicated to observing 50 to 100 fields in the sky, far away from the band of the Milky Way, each containing about 1000 galaxies. Powerful computers compare the images of the same field from different months and search for changes; typically, about two dozen supernovae are detected in this way.

Several large telescopes in Chile, Hawaii, Arizona, and the Canary Islands, as well as the Hubble Space Telescope, can then follow the changes in brightness after the supernova explosions and determine the redshift in the spectra of the parent galaxies. In addition to the modern detectors and telescopes needed for this project, the Internet has proved invaluable: without it to move the wealth of data around the world, this supernova search program would be impossible. One result of this program shows that the spectra of very distant supernovae are identical to those nearby, which indicates that the physics of these stellar explosions has not changed over the last 5 billion years. By January 1998, 40 complete supernova observations were available, and their implications were significant. The points began to deviate notably from the curve for the critical density, in a direction that indicates a density much below the critical density. Therefore, the favorite model of many theoreticians—a flat universe without a cosmological constant and at critical density—was ruled out with a probability of 99 percent, and the perpetual expansion of the universe was almost proved. Such clear statements constitute a completely new situation in observational cosmology.

There is still some latitude for the various parameters, but a lower than critical density and a positive cosmological constant have become highly likely. One study by itself is not sufficient to predict the future of the universe reliably, but there are others under way. Another program that searches for type Ia supernovae in distant galaxies has yielded many good results, among them the most distant supernova, within a galaxy of redshift 0.97. Again, the conclusion indicates a less than critical density and a positive cosmological constant. The data may even indicate that the expansion rate of the universe is currently increasing. This effect of a positive cosmological constant had not been expected widely, and rumors of a mysterious "antigravity" made their

Three distant supernovae tell the future of the universe. These stellar explosions occurred in 1997 in galaxies with redshifts of 0.50, 0.44, and 0.97 (right)—at that time, a distance record for supernova observations. (High-z Supernova Search Team and NASA)

way into the media. Even this case, however, is "permitted" by Einstein's theory of general relativity.

Completely different methods of determining the geometry of the universe now yield consistent density values around 20 percent or less of the critical density. In January 1998, at the meeting of the American Astronomical Society in Washington, D.C., the cosmologists themselves were surprised at how well the various values agreed with one another: even a year earlier, any value for the density of the universe still seemed possible. Observations continue to con-

strain these parameters further, and the Hubble Space Telescope remains in high demand, particularly to determine the precise brightnesses of very distant supernovae. NICMOS, the new infrared camera, will be used to observe the most distant, reddest supernovae, thus helping scientists to gain more and more accurate values for the density of the universe and for the cosmological constant. These possibilities have lifted the field of observational cosmology to a fundamentally new level of quality and accuracy. Only to the question of questions—why are the values of the various parameters what they are, and not something else (zero, for instance)? — can Hubble not provide an answer. For now.

Searching for the Building Blocks of Galaxies

The search for the origin of today's galaxies has always been one of the central tasks of astrophysics. The Hubble Deep Field had already demonstrated the change in the appearance of the universe over the last several billion years. Another long exposure from HST provided additional input for the developing theories. The image contained 18 irregular systems of stars, smaller than full-grown galaxies but still experiencing active stellar formation. All of them are located at roughly the same distance (about 11 billion light years) and are very close together, so that several future mergers appear likely. Are these "pregalactic blobs," as they have been called by their discoverer, the building blocks of more recent galaxies and the proof that galaxies formed not at once, but from smaller fragments? Aside from the Hubble Deep Field, this image was one of Hubble's longest exposures: WFPC2 had collected photons from a region in the northern constellation Hercules for 67 orbits.

Only when more celestial fields have been studied to similar depths will the characteristic image of the early universe emerge with more certainty. But this deep image is certainly not unusual, for it contains as many faint blue galaxies, the precursor of current galaxies, as other fields. The 18 blobs, detected by their redshifted ultraviolet radiation coming mainly from hydrogen, are located in an area only 2 million light years in diameter. Never before have so many systems undergoing star formation been seen so close together. Each blob consists of about a billion stars and has a diameter of 2000 to 3000 light years.

The continuing process of star formation is indicated by the presence of young blue stars and glowing gases. Repeated mergers of these objects could produce the thick central region (the so-called bulge) of a typical recent galaxy; the bulge of our own Milky Way has a diameter of about 8000 light years. In the early universe, the rate of mergers of galaxies was much higher than it is today; at least four of the blobs evidence such merging activities in a binary structure in their centers.

These blobs can be seen as evidence for the "bottom-up" model of galaxy formation and the theory of cold-dark-matter cosmology. In this model, the universe is initially full of "cold" dark matter, moving much slower than the speed of light. Structure emerges from the bottom to the top, or from small to large. First, small stellar systems—such as star clusters and small galaxies—emerge. Next, they combine into larger galaxies like our own Milky Way. Finally, the galaxies form clusters and superclusters. This sequence would also explain the sheetlike and filamentary structure of the current universe. If the initial dark matter were "hot," however, and consisted mainly of particles near the speed of light (neutrinos with minimal mass, for instance), then the large structures would emerge first, fragmenting into galaxies only later. The supporters of the cold-dark-matter cosmology see the blobs in the various Hubble images as confirmation of their theory, because these would be the building blocks from which galaxies form.

The stars emerging in the blobs would later form the bulges of galaxies. The remaining gas of the blobs would become the galactic disk, where additional stars and also spiral arms would develop later. Dwarf galaxies like the Magellanic Clouds could be blobs that missed merging into larger galaxies. Globular star clusters would be initial congregations of stars that did not even make it to blobs. But an alternative scenario cannot be ruled out: it might be that the blobs are only brighter concentrations in a large, diffuse precursor of a galaxy that is too faint to be seen in the current Hubble images. Even a combination of these scenarios might be possible. Whichever scenario turns out to be true, the Hubble Space Telescope made it feasible for the first time to observe objects at the time of the formation of galaxies with sufficient angular resolution to help answer these fundamental questions. The large infrared satellite SIRTF, after its planned launch in 2001, and later the Next Generation Space Telescope (NGST), should be able to provide more information in this area.

The Hubble Space Telescope had been looking for the precursors of today's galaxies since its launch. In 1995, the magnitude of the changes in the universe over billions of years became apparent when the analysis of the "medium-deep survey," a creative random observing program, was completed. Whenever an object was observed for any length of time with one of the other scientific instruments, WFPC2 was used to take images of the vicinity of the original target. On several occasions, fairly long exposure times were accumulated. In a way, this project was a precursor of the Hubble Deep Field. The medium-deep survey images revealed that a few billion years ago, irregular and faint blue galaxies were much more numerous than they are today, when large spiral and elliptical galaxies dominate. Ten times as many irregular galaxies as expected appeared in a particular field with a fairly long exposure time. The later development of these faint blue galaxies, however, cannot be deduced from these images: Did they merge into the current galaxies, or did they simply fade away?

Building blocks of the universe at a sixth of its current age? Eighteen young galaxies or their precursors were found in this deep Hubble image. The objects are 2000 to 3000 light years in diameter, much smaller than present-day galaxies. They are also close enough together to collide within the next few billion years. (Windhorst & Pascarelle and NASA)

Colliding Galaxies

For decades, astronomical textbooks had classified galaxies as three basic types: elliptical, spiral (with and without a central bar), and irregular. It was assumed that all types of galaxies had formed in the distant past with those shapes and that they had kept those shapes up to the present. Considering that the distances between galaxies are only about 100 times larger than their diameters and that they move through space with relative velocities of several hundred kilometers per second, "cosmic traffic accidents" may occur fairly frequently. What happens during such a collision of galaxies? Collisions between the stars in galaxies are very rare because their distances apart are millions of times larger than their diameters. Even in a galactic collision, the individual stars of the galaxies involved will collide only very rarely. Gas clouds in the colliding galaxies, however, are compressed and heated in the collision, which leads to very active star formation in these areas. The observation of regions of surprisingly strong star formation activity in a galaxy is therefore always an indication that this galaxy may have been involved in a collision.

Galaxies with high star formation rates, called starburst galaxies, have indeed been found in large numbers. In many cases, the starburst can be attributed directly to a collision, and sometimes to a close encounter or near miss with another galaxy.

The investigation of galactic interactions started surprisingly early. During the 1940s, the Swedish astronomer Erik Holmberg built two model galaxies using 100 light bulbs for each. The computers of that era were not powerful enough to calculate the gravitational field for a sufficiently large number of model stars in a galaxy, so Holmberg simply measured the equivalent "illumination" by using photocells at the locations of the light bulbs. The resulting movement of the model stars could then be determined. During the 1950s, the Swiss astrophysicist Fritz Zwicky used the 5-meter telescope on Palomar Mountain to photograph interacting galaxies. He found tails and antennalike structures emanating from the interacting galaxies, reminiscent of Holmberg's earlier calculated results.

In the 1960s, the American astronomer Halton Arp published an atlas of peculiar galaxies: a compendium of colliding galaxies and bizarre remnants from such collisions. The universe appeared to be a violent place, full of galactic collisions, cannibalism, and mergers. A decade later, Alar and Juri Toomre performed the first modern computer simulations of galactic collisions; these simulations explained the distortions and deformations observed by Zwicky and Arp. A fairly large fraction of today's galaxies may actually be remnants of galactic mergers—and these should be mainly elliptical galaxies. One problem remained. Spiral galaxies include large amounts of dust and gas, whereas ellipticals contain many globular star clusters. For instance, our own Milky Way—a spiral galaxy—has only about 150 globular clusters, whereas an elliptical galaxy of comparable brightness would have about 600. The only reasonable explanation is that many globular clusters formed during the bursts of star formation resulting from a collision. But astrophysicists had long believed that globular clusters contain the *oldest* stars in the universe. Indeed, globular clusters had been used to determine the lower limit for the age of the universe.

HST has now observed several colliding galaxies and investigated the corresponding globular clusters to estimate their approximate age from their colors. These observations allow the stellar and cluster formation rates during and after a collision to be examined.

The most spectacular Hubble image of a pair of colliding galaxies is the view of the central part of the Antenna Galaxies NGC 4038 and 4039—the prototype of a galactic collision. In addition to the reddish galaxy cores, we can see chains and groups of more than 1000 bluish star clusters. With an age of about 50 million years, these are newly formed future globular clusters. Although it is certainly possible that all elliptical galaxies have been formed from merging spirals, the Hubble observations cannot prove it.

There are other noteworthy examples of galactic collisions. The Cartwheel Galaxy resulted from a head-on collision of a spiral galaxy and another small galaxy, producing a remarkable spoke structure. The yellow core of the Cartwheel Galaxy shows a network of dust lanes, but the large star-forming regions are missing. The central ring, however, is surrounded by a group of gas clouds that look like giant comets;

the white "heads" have diameters of several hundred light years, and the blue tails are 1000 to 5000 light years long. Large amounts of hydrogen are expected in the heads. These gas clouds could have been ejected during the collision 200 million years ago and could now be falling back into the galaxy. As they hit the ring, a shock structure develops. While many details of the Cartwheel Galaxy remain enigmatic, the galaxy that caused the collision is known. With the Very Large Array (VLA) radio telescope in New Mexico, a trace of neutral hydrogen has been detected, pointing to the culprit.

Another example is Arp 220, two colliding spiral galaxies that caused an active burst of star formation 100 times larger than the current star formation rate in our own galaxy. NICMOS observations helped us to look deeper into this dusty system.

The effects of a galactic collision on the innermost parts of the participating galaxies can be seen in great detail in Centaurus A. Located at a distance of about 10 million light years, it is the closest active galaxy and radio galaxy. Centaurus A is an elliptical galaxy with a prominent dust lane, and by 1954 it was already suspected of being the result of the merger of an elliptical galaxy with a small spiral galaxy. Hubble's WFPC2 camera has probed this dust lane in much greater detail than ever before. It has revealed star clusters with young blue stars located at the edge of the dust layer. Apparently, the collision let to violent star formation in this case as well. NICMOS images provide a fascinating view into the

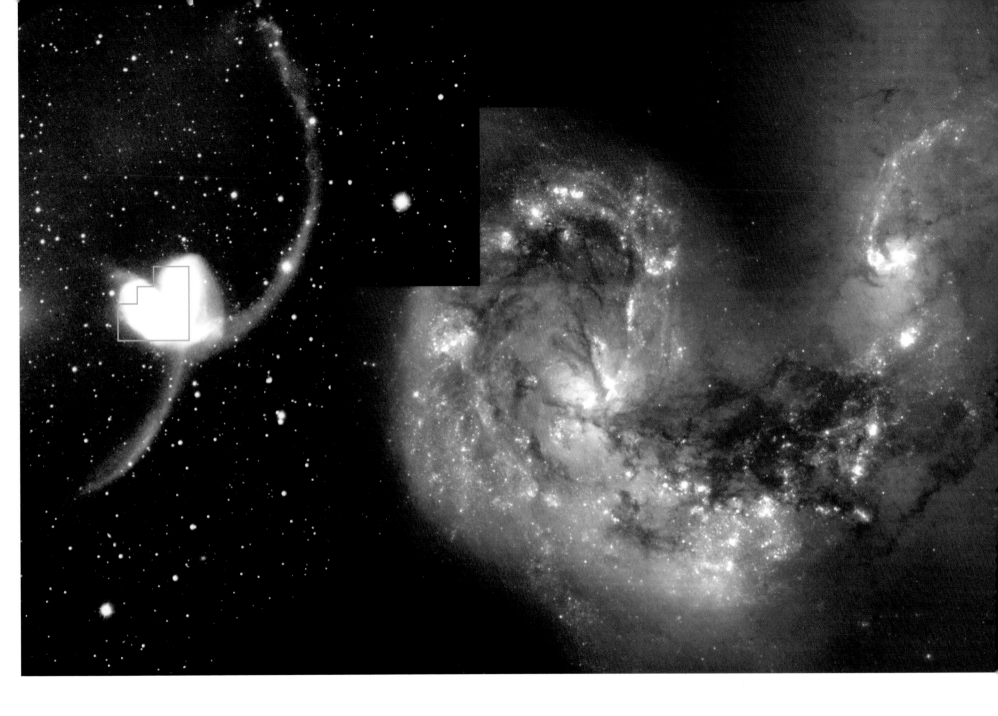

Traces of a cosmic collision. The entire Antenna Galaxy, seen in a ground-based image on the left, is the result of the interaction of NGC 4038 and NGC 4039. The Hubble image of the central region of the two galaxies is shown on the right. The extensive spiral arms reflect the burst of star formation after the collision. (Whitmore and NASA)

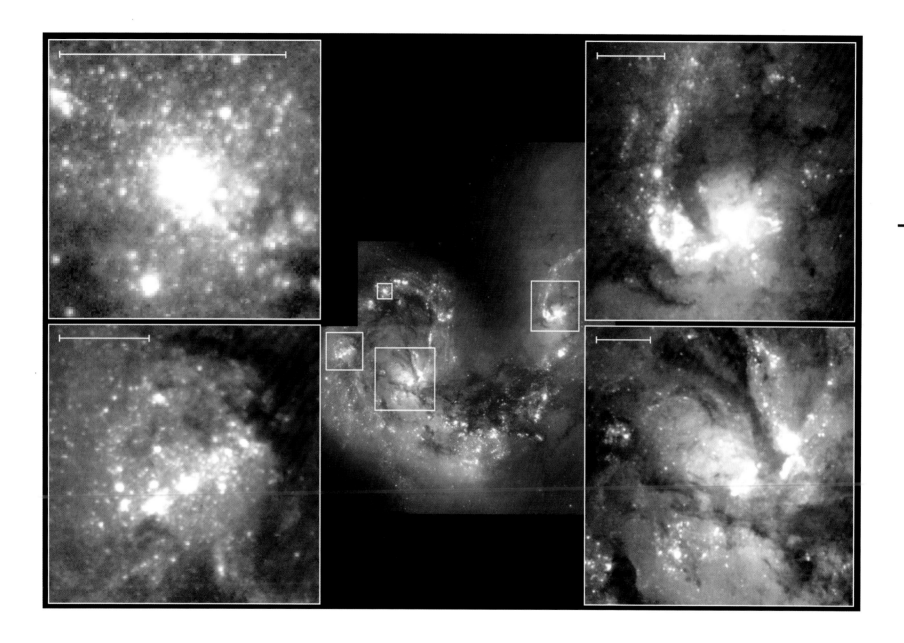

Star formation in the Antenna Galaxy. Brilliantly radiating star clusters have developed as a result of the collision of two spiral galaxies. (Whitmore and NASA)

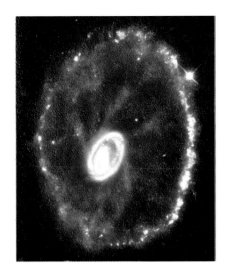

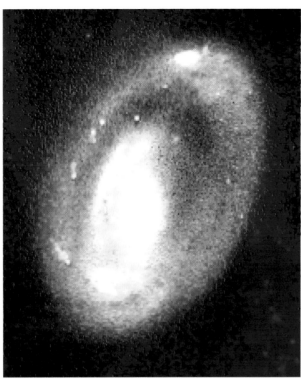

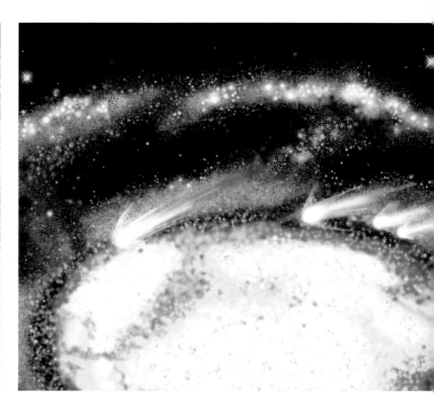

interior of Centaurus A, from where strong radio jets emerge. The camera peered past the dust to discover a tilted disk of hot gas at the galaxy's center. This disk apparently provides the "feeding" of the galaxy's central engine, a suspected supermassive black hole that may be a billion times the mass of our Sun.

Stars between Galaxies

Not only do galactic collisions cause fireworks in the interiors of the galaxies involved, but they also cause stars to be ejected from the galaxies into intergalactic space. For the past 50 years, it has been suspected that intergalactic stars exist in galaxy clusters. These stars were torn from their parent system during the formation of the galaxy or during later interactions with other galaxies. The presence of a diffuse glow in the interior of galaxy clusters and the discovery of planetary nebulae—the glowing remnants of extinct stars—between the galaxies of the Virgo cluster appear to confirm this hypothesis. The mass contained in these intergalactic stars could not be determined accurately until Hubble was able

to provide some answers. A deep image of a field devoid of galaxies, but within the Virgo cluster, was compared with the Hubble Deep Field, which provided the cosmic "background." The Virgo cluster field showed an excess of about 630 stars! Only about 20 of them could be foreground stars within our own galaxy; the remaining 600 or so must be red giant stars in the intergalactic space of the Virgo cluster.

In addition to the bright red giants, there should also be 10,000 times as many faint stars still on the main sequence of stellar evolution; but these are too faint to be seen by Hubble. Adding the masses of the visible and suspected stars shows that 4 to 12 percent of the total cluster mass is contained in the intergalactic stars of the Virgo cluster, and extrapolations based on the planetary nebulae yield even higher percentages. The intergalactic red giants confirm the old hypothesis that during close encounters of galaxies—which should be fairly frequent in a galaxy cluster—stars can be torn from their parent systems. These stars could be used as standard candles for measuring distances; they could also help to determine the distribution of dark matter in a cluster. But it is

Supersonic "comets" in the Cartwheel Galaxy, the result of the collision of a small galaxy with a large one. In the inner region, clumps of gas that resemble comets are being discovered, with "heads" a few hundred light years in diameter and "tails" that extend over several thousand light years. These shapes indicate that fast-moving material has plowed into slower gas. The diagram at the right clarifies this effect. (Struck et al. and NASA; Gitlin) still uncertain whether intergalactic stars are a cosmic curiosity — a by-product of celestial mechanics in galaxy clusters — or an important phenomenon in the universe.

In some galactic interactions, a larger galaxy can steal the star clusters from a smaller companion, a phenomenon that has been observed with HST. The "thief" is the giant elliptical galaxy M87 in the Virgo cluster. Based on WFPC2 images, several star clusters were found in M87 that originally belonged to its satellite galaxies NGC 4486B and NGC 4478. M87 can be convicted of stellar theft only by circumstantial evidence, however: the satellite galaxies contain significantly fewer globular clusters in their outer regions than comparable galaxies at greater distances from any disturbing large system. In addition, the small galaxies are much more compact than normal, which indicates that they must have lost the stars of their exterior regions, as well as their globular clusters, to M87. It has long been suspected that M87 resulted from the merger of several smaller galaxies — but even after the merger, it seems to have accumulated clusters and stars from surrounding galaxies.

NOAO

HST

A burst of star formation in Centaurus A. A dense dust lane is evidence of a cosmic collision; the bright blue clusters of young stars resulted from the collision. The remainder of the galaxy radiates in the reddish light of older giant and dwarf stars. The dust disk is seen almost edge-on, and its thickness indicates its relatively young age. (Schreier et al. and NASA)

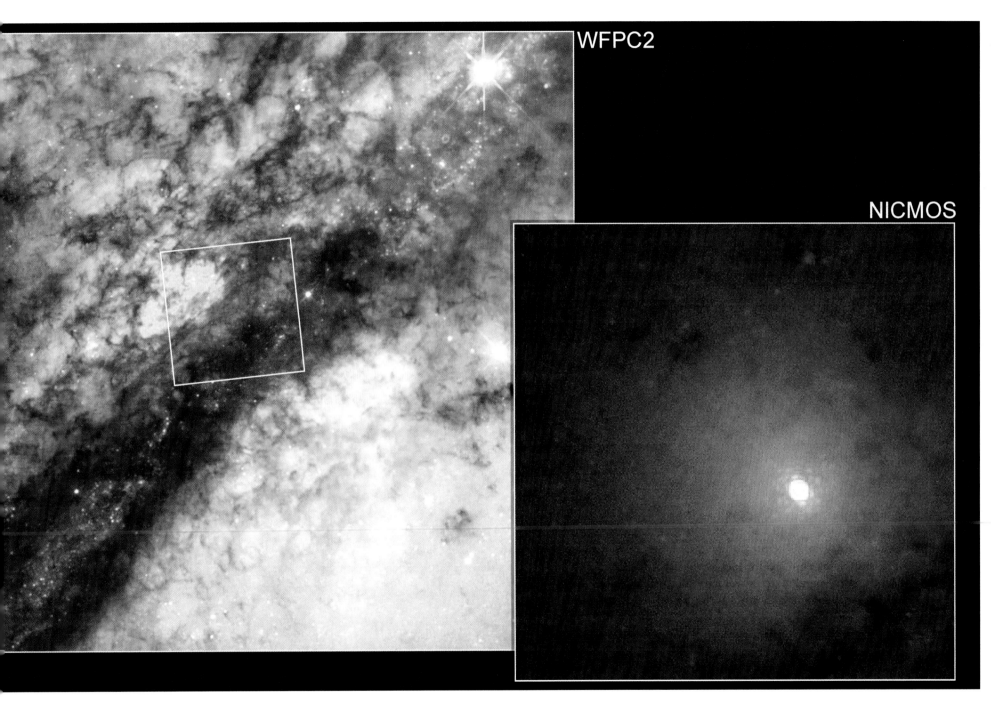

WFPC2

NICMOS

The interior of Centaurus A is hidden in visible light — but Hubble's NICMOS infrared camera can see it. The image on the right was taken at a wavelength of 1.9 microns and reveals a large disk of hot gas, 130 light years in diameter, surrounding the central engine of Centaurus A. The red bubbles are luminous gas clouds, energized by the strong radiation from the central engine. (Schreier et al. and NASA)

Quasars: Beacons at the Beginning of Time

For a long time, these objects had been considered faint stars, similar to millions of others. The brightness of some of them varied unpredictably, and these were cataloged as variable stars. After radio astronomy had come of age and was able to determine the positions of radio sources in the sky with greater accuracy, it turned out that several radio sources coincided with faint blue "stars." The peculiar term *quasar*, short for quasi-stellar radio source, was given to these objects. One optical identification involved the radio source 3C273. When the visible light from this object was analyzed, the spectrum appeared abnormal: a number of bright and broad lines were located at wavelengths with no known equivalent in the spectra of normal stars. During the early 1960s, an astronomer at the Palomar Mountain Observatory, Maarten Schmidt, discovered that these lines belonged to certain chemical elements, but that they were shifted to the red by incredible amounts. The redshift corresponded to a velocity of 30,000 kilometers per second, a tenth of the speed of light! Nothing like this had ever been seen before. Should this redshift really be interpreted as cosmological—that is, based on the expansion of the universe—which would make quasars very distant objects? Or did another physical effect cause the redshift?

It was quite some time before the cosmological interpretation of the redshift of quasars—and their consequent large distances away from us—was universally accepted. Given their brightness in the sky and how distant they are, quasars must be incredibly luminous. They would have to emit several thousand times more energy than normal galaxies in our vicinity. Even more surprising was the realization that the relatively fast brightness variations in some quasars indicate that the energy is released in an area the size of our solar system. Today, the most popular theory for the "central engine" of a quasar postulates a massive black hole in the center of a galaxy, feeding on large amounts of gas and large numbers of stars. The matter falling into the black hole forms a very hot disk of matter around it; this disk can provide an amount of energy that explains the observed luminosities. Early in the evolution of the universe, galaxies were rich in stars and gas; today's galaxies contain much fewer stars and much less gas. Once the gas and stars in the vicinity of a black hole have been used up, the black hole, still located at the center of the galaxy like a spider in its web, is being "starved" and indicates its presence only indirectly by its large mass.

The Hubble Space Telescope has provided several important clues to understanding quasars. One question was: are quasars really located in the centers of galaxies? Ground-based telescopes had found indications of faint nebulosity around a few quasars, but the limited image quality and lack of sufficient resolution did not permit the nature of the galaxies to be determined. Are they spirals? Ellipticals? Galaxies in collision, thus providing new supplies of matter to their "central engines"? Even Hubble had a hard time providing definitive answers. Early in 1995, a

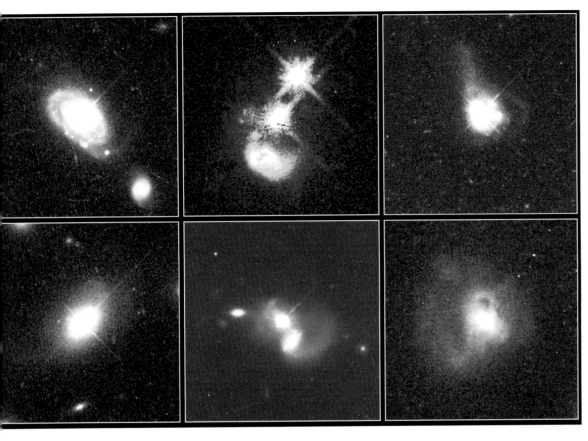

The homes of quasars: a variety of galaxies from normal (left column) to colliding (center) and peculiar (right). See the text for detailed descriptions. (Bahcall & Disney and NASA)

team of scientists reported that, based on Hubble observations, the classical concept of quasars had been turned upside down—a large fraction of quasars had been found "naked," without a surrounding host galaxy. This conclusion, however, was based on a series of images to which extreme amounts of image processing had been applied. The results were suspect, at least in the eyes of other scientists, and in fact they were retracted later that year.

The connection between quasars and galaxies was still not clear. Extremely powerful quasars, with luminosities of 100 to 1000 times those of normal galaxies, exist in a remarkably large number of different types of galaxies, some of them just beginning to undergo collisions, others looking perfectly "normal." Are there perhaps several different mechanisms that initiate quasars? The left column of images on page 77 shows quasars in a normal spiral galaxy (top, PG 0052+251) and in a normal elliptical galaxy (bottom, PHL 909).

The other four images show quasars in colliding or merging galaxies. In the top center image, two galaxies collide with a velocity of about 450 kilometers per second. The debris from this collision may be fueling quasar IRAS 04505-2958, which is 3 billion light years from Earth. Astronomers believe that a galaxy plunged vertically through the plane of a spiral galaxy, ripping out its core and leaving the spiral ring (at the bottom of the picture). The core, the bright object at the center of the image (the object above is a foreground star that has nothing to do with the quasar), lies in front of the quasar. Surrounding the core are star-forming regions.

In the bottom center image, Hubble has captured quasar PG 1012+008, merging with a bright galaxy (the object just below the quasar). The swirling wisps of dust and gas surrounding the quasar and the galaxy provide strong evidence for an interaction between them. The compact galaxy on the left of the quasar also may be beginning to merge with the quasar. In the top right image, Hubble has observed a tidal tail of dust and gas beneath quasar 0316-346. The peculiarly shaped tail suggests that the host galaxy has interacted with a passing galaxy that is not in the image. At the bottom right, we see evidence of a dance between two merging galaxies. The galaxies may have orbited each other several times before merging, leaving distinct loops of glowing gas around

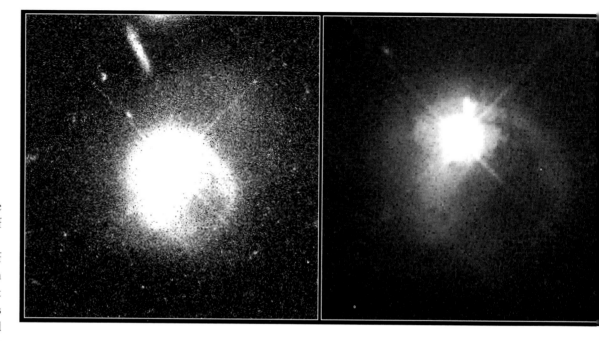

quasar IRAS 13218+0552. The elongated core in the center of the image may comprise the two nuclei of the merging galaxies.

The image on page 78 shows another case of a quasar associated with huge, thin tidal arms. In the right-hand panel, the same image is shown at a different contrast level. Only 11,000 light years separate the quasar and the companion galaxy (located just above the quasar). This galaxy is similar in size and brightness to the Large Magellanic Cloud near our Milky Way. The galaxy is closer to the quasar's center than our Sun is to the center of our galaxy. The quasar and galaxy are drawn together by strong gravitational forces. Eventually, the galaxy will fall into the quasar, potentially giving rise to more brightness fluctuations.

All these are interesting special cases—but what can statistics tell us? Teams of Hubble observers do not fully agree. One group found indications of interacting galaxies in 11 of 15 quasars; another team found such indications in only ten of twenty cases, with the remaining galaxies appearing "normal." Both groups, however, agree on several other findings:

- Most quasars are located in luminous spiral or elliptical galaxies; Hubble observations permitted this classification for the first time.
- Galaxy collisions and interactions play an important role in inducing the quasar phenomenon, but other mechanisms may also occur in undisturbed, normal galaxies.

- "Radio-quiet" quasars without detectable radio emission are located not only in spiral galaxies, as had long been believed, but also in ellipticals.

The results of another large Hubble observing program of 33 quasars, presented in March 1998, indicates that giant elliptical galaxies are the most common host galaxies for quasars. Further results will come from Hubble's new NICMOS infrared camera, and particularly from the Advanced Camera for Surveys, once it is installed during Servicing Mission 3 in 2000. This instrument will have a built-in coronograph, which is an optical system that can mask out the light from the quasar, so that the surrounding host galaxy can be studied in more detail.

There is also a question about the duration and frequency of the quasar phenomenon: is it common but short-lived—leading to the expectation that a burnt-out quasar exists in almost every galaxy—or is it long-lived but rare? The nature of the central engine in which the quasar processes occur may

A quasar in merging galaxies. The identical image of the quasar PKS 2349 is shown twice at different contrast levels to display the outer regions of the galaxy and the area closer to the quasar. (Bahcall and NASA)

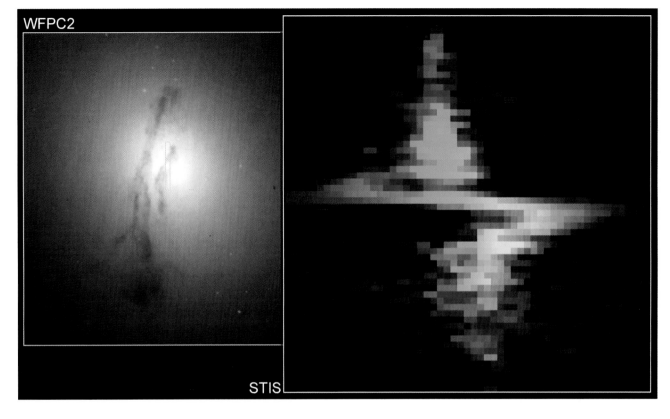

The signature of a black hole? The colorful zigzag line on the right is one of the first measurements taken with Hubble's new STIS spectrograph. Its entrance slit extended across the galaxy M84 in the Virgo cluster. The extension of the line toward left and right indicates increasing velocity of the gas. The velocity increases near the center of the galaxy to about 400 kilometers per second, indicating a central object of about 300 million solar masses. (Bower & Green and NASA)

also merit some further investigation. Are quasars really supermassive black holes, or can other lesser known models explain the phenomenon better? One such model involves a "burning disk," a disk-shaped giant star in the center of a galaxy that would provide the required energy. This model would make apparently normal and undisturbed host galaxies easier to explain: a gravitational disturbance caused by a collision and feeding the central engine would take long enough to migrate toward the center of the host galaxy that the outer regions would have had time to regenerate.

Unfortunately, quasars are too far away to permit Hubble to observe the details of the central mechanisms directly. "Relatives" of quasars, however, exist even in the cosmic present, and therefore closer to us, and they allow us to gain more information about these processes.

Active Galaxies: Nearby Mini-Quasars

When the American astronomer Carl Seyfert observed galaxies with bright cores as early as 1943, hardly anybody took note of his discovery. Later, Seyfert became director of the Dyer Observatory in Nashville, Tennessee, where he had only small telescopes at his disposal. He died at a young age in a car accident and did not get to witness the increasing importance of "his" galaxies.

NGC 4151 is the brightest Seyfert galaxy in the sky. It is expected that, similar to quasars, the activity in the cores of these systems is caused by the accretion of matter into a massive black hole. The energy output of a Seyfert galaxy is much lower than that of a quasar, however. If quasars are the mighty dinosaurs of prehistoric times, then the Seyfert galaxies are their lizard descendants. Seyfert galaxies do not appear just as blinding points of light but instead reveal their inner workings to the observer. The HST image of NGC 4151, for instance, shows two luminous gas cones consisting of many concentrations or clumps of gas. A radiation source in the center of the galaxy illuminates this gas. Hubble's new STIS spectrograph can analyze the velocities of hundreds of these gas clumps simultaneously. Each gas clump emits emission lines of several chemical elements, and their wavelengths are shifted in correspondence with their individual velocities. Analysis of these observations indicates that most of the gas clumps stream outward. Sometimes Seyfert galaxies are so rich in dust that the central engine is totally obscured—until an infrared camera like NICMOS lifts that veil. In several other cases, Hubble has captured surprising details in the vicinity of the central engine even in visible light. In NGC 6251, for instance, we see the bright ultraviolet light escaping from the central region on only one side of the central dust disk, which is warped like the brim of a hat. The most impressive case may be NGC 4261, where the dust disk is perfectly visible. Inside that dust disk lies the gas disk; its large rotational velocity allows the mass of the central object to be determined: about a billion solar masses.

It is interesting to hear that a supermassive black hole may have been detected in this or that galaxy, but other questions are even more interesting. How common are black holes in the centers of galaxies? Is their mass related to the mass of the host galaxy? To answer these questions systematically, 27 nearby galaxies were examined with the Canada-France-Hawaii telescope on Mauna Kea, Hawaii, and evidence of black holes was found in three of the normal galaxies. The evidence is the high velocity of stars circling the centers of the galaxies. The galaxy NGC 3379 (or M105) contains 50 million solar masses, NGC 3377 contains 100 million, and NGC 4486B contains 500 million, concentrated in a comparatively small space in their centers. Such massive concentrations of matter that emit very little, if any, light are normally interpreted as black holes. Several trends begin to appear:

- Almost every large galaxy appears to harbor a

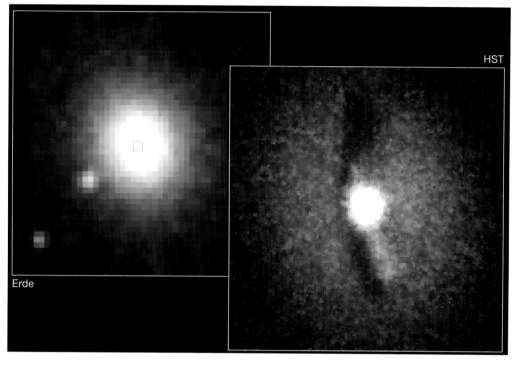

The core region of the galaxy NGC 6251 in a combination of WFPC2 and FOC images. In visible light, a dark dust disk can be seen, whereas the ultraviolet shows increased brightness on one side; this means that the disk must be warped like the brim of a hat. The hot point source in the center represents the radiation from the central engine, a suspected black hole. (Crane and NASA)

supermassive black hole.

- The mass of the black hole may be correlated to the mass of the galaxy.
- The numbers and masses of quasars fit the hypothesis that in the centers of most galaxies, a quasar has been active at some time, leaving behind a massive, but inactive, black hole.

The recent universe, therefore, would seem to be full of quasar "fossils"—and that begs the question, why don't all of them exhibit activity in a brightly radiating center, as Seyfert galaxies do? Apparently, this has to do with the inadequate "feeding" of the central engine: matter from the bulge or disk of the galaxy must be available for activity to occur. A typical Seyfert galaxy needs only about one solar mass per year to sustain its activity, but that amount has to be transported toward the center somehow. Dust could play an important part in this transportation, as recent Hubble images of two galaxies suggest. NGC 1667 contains dust that moves toward the center in a spiraling motion. A similar effect is seen in NGC 3982. The details of this mechanism remain unclear, but it provides an alternative to the only other known way to transport matter into the galactic center. The latter mechanism involves complicated trajectories of stars and the formation of a central bar of matter in spiral galaxies. The next step will be to compare active galaxies with inactive ones. How are they different?

Gravitational Lenses: Hubble's Telephoto Lens

For his general theory of relativity, Einstein developed the test case of curvature of space caused by a sufficiently large mass. In the case of the Sun, the light path from stars appearing close to it should be bent, so that the stars would seem to move a tiny amount away from the Sun. And this effect was indeed measured during solar eclipses. In the 1930s, the Russian physicist Chwolson published a paper in which he stated that the curvature of space not only would cause a spatial distortion but also could lead to amplification of the light. Einstein and Zwicky made similar predictions. But nobody believed that such "minimal" effects could ever be observed in the universe. What would happen if the gravitational lens were caused not by a star but by a massive galaxy or even a cluster of galaxies? This should enhance the effect considerably. Several decades passed before the first cases of gravitational lenses were observed. In 1979, a quasar "lensed" into a pair of points by a foreground galaxy was discovered. Since 1988, several arcs and spots around galaxy clusters have been observed that turned out to be "lensed" background galaxies.

Because the shape of these arcs and their distance apart can be modeled mathematically, the mass causing the gravitational lens can be determined. This determination is independent of the nature of the mass; that is, whether it contains dark or luminous matter. The galaxy cluster CL0024+1654 acts as a gravitational lens on an even more distant cluster, producing eight distinct images of the distant cluster. These images allowed an accurate map of the distribution of mass in CL0024+1654 to be derived. Astrophysicists needed more than a million complex model calculations to arrive at the mass distribution causing the observed images. The luminous galaxies contribute less than half a percent to the total mass of the cluster, gas and dust about 10 percent—and fully 90 percent of the cluster mass is dark and of an unknown nature. This mysterious dark matter is spread equally over the entire cluster. Another result indicates that a fraction of the galaxies have less dark matter in their halos than others; they may have lost it in collisions.

The typical effect of a gravitational lens distorts the image of the background object into several arcs. Impressive examples include the galaxy cluster CL0024+1654, "imaging" a twice as distant blue galaxy five times, and RX J1347.5-1145, the galaxy cluster with the largest x-ray luminosity in the sky and with one of the largest known masses. Hubble's STIS images show several arcs of a lensed, more distant background object.

If the background object and the lensing object were perfectly aligned with regard to the observer, then the lensed object would appear as a perfect circle. Einstein described this case in one of his publications, but he did not believe that it would ever be found in nature. Nature should never be underestimated: a British radio telescope discovered a nearly perfect ring that was later confirmed by infrared images from HST. The radio image of

An entire cluster of galaxies acts as a gravitational lens. The combined gravitational field of the yellow elliptical and spiral galaxies has dramatically distorted the image of a more distant blue galaxy into five separate parts. The galaxy is about twice as far away as the cluster CL0024+1654; the galaxy's spiral shape has been considerably changed by the lensing effect, but several details are still visible. Its clumpiness, for instance, indicates its youth and active star formation. (Colley et al. and NASA)

B1938+666 looks like a typical gravitationally lensed image. The small radio source in the center of the galaxy is distorted into one of the typical arcs; it is not quite aligned with the lensing foreground object. But the NICMOS image reveals a nearly perfect ring: the complete galaxy, not just its center, is visible in the infrared, so that part of it lies exactly behind the close galaxy; light from this part of the galaxy forms the ring. The ring is not totally uniform; the brighter central region of the galaxy is distorted into a brighter arc, superimposed onto the top third of the complete ring.

"Einstein crosses" are another special case of gravitational lenses: a distant quasar is located almost exactly behind a foreground galaxy that produces several quasar images surrounding the galaxy in the form of a small cross. These cases were considered at least as exotic as complete rings, and only two objects of this type were known. The medium-deep survey carried out with HST has increased their number dramatically. Two more crosses were discovered in images taken over a total area only the size of the full Moon. Extrapolating from these results, the entire sky would then contain about half a million such gravitational lenses. Once enough gravitational lenses have been observed, cosmologically relevant statements can be made—for instance, about the average density of the universe, and therefore about its future. Even the first few Hubble results indicated, albeit with a number of reservations, that the average density of the universe is less than the critical

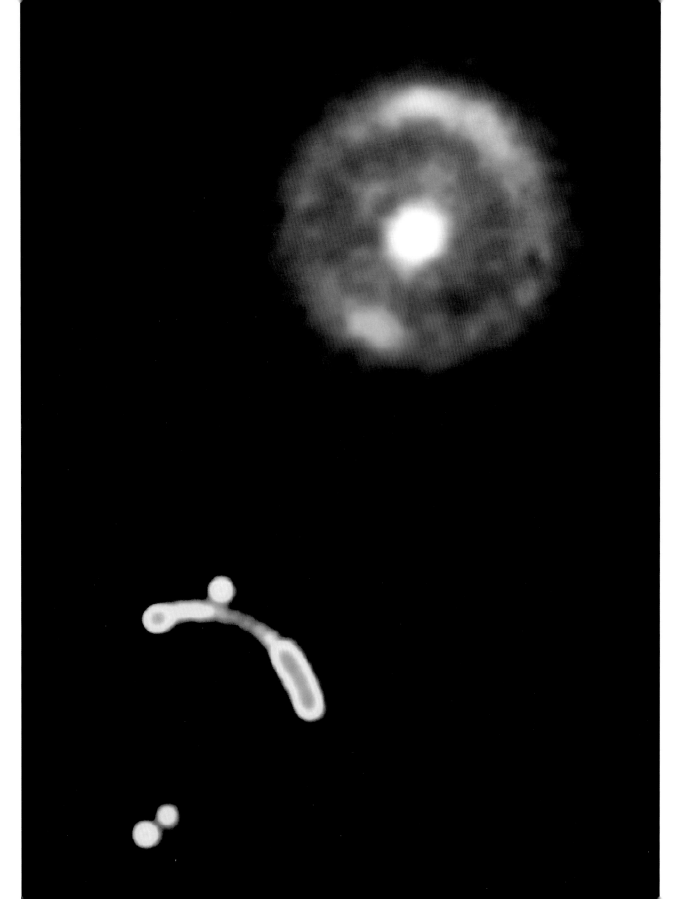

An almost perfect Einstein ring is produced by this gravitational lens. Although the radio image (bottom) shows only part of the ring formed by a foreground galaxy from a distant source, Hubble's NICMOS image (top) makes the entire ring visible. (King and NASA)

density, which confirms other findings based on completely different approaches.

Gravitational lenses can be used as auxiliary telescopes of cosmic proportions. Although they may distort distant objects, they also can brighten their images, making them more clearly visible. The presumed "brightest galaxy in the universe" of the early 1990s, for instance, was proved by a later Hubble observation to be a gravitationally lensed image. The gravitational lens allowed the image to appear brighter by a factor of 30. The Hubble image, with its superior resolution, showed the characteristic arc structure that is only produced in gravitational lenses. Several scientific papers about the presumed monster galaxy had been published prematurely. But astronomers usually appreciate gravitational lenses, particularly if they are located in front of a distant galaxy. Luckily, the distances of both the lensed and the lensing object can still be determined without any problem. The light of the distorted object still carries the redshift information, which is not altered in the lensing process.

The first "telescopic" use of a gravitational lens dates back to 1995 and involved the galaxy cluster Abell 2218. An arc in the image has been enhanced by the lens by a factor of 15, as model calculations showed. This image encouraged the closer investigation of the object that caused the arc. The spectrum reveals that it is a galaxy with a redshift of 2.5; the lensing cluster, with a redshift of only 0.18, is significantly closer. This spectrum also indicates the presence of strong star formation activity.

In 1997, a gravitational lens also established a new cosmic distance record, which stood for only a few months. This was the first case where the distorted image was mathematically reconstructed to show the true shape of the object. The galaxy cluster CL1358+62 is located at a distance of about 5 billion light years. In its vicinity, the distorted images of an even more distant galaxy are visible. The strangely shaped object first drew attention by its deep red color. Hubble observations, with their superior resolution, revealed some structure. And one of the 10-meter Keck telescopes on Mauna Kea, Hawaii, succeeded in obtaining a spectrum of this galaxy. An emission line found at a wavelength of 720 nanometers could be identified as the hydrogen line Lyman alpha; its rest wavelength, as measured in the laboratory, is 122 nanometers. Therefore, this object (and its close companion, discovered later) exhibit a redshift of 4.92. This number exceeded that of the long-time distance record-holder, a quasar with a redshift of 4.90; a more distant galaxy was then discovered in 1998. The temporary distance record is not the most important feature; rather, the gravitational field of the lensing cluster was modeled, the lensing effects were calculated, and the observed distorted image was reconstructed to show the true shape of the galaxy. In a manner of speaking, Hubble looked through a fairly inferior but still usable telephoto lens. The resulting reconstructed image is

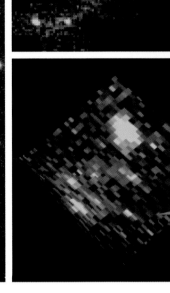

5 to 10 times more detailed than it would be without the lens.

The brightest spot of light is barely resolved; its diameter is about 800 light years, and it is more luminous than star-forming regions in starburst galaxies. It is possible that we are witnessing the formation of the core region of the galaxy, the galactic bulge. Several other starburst regions, each about 700 light years in diameter, are visible, distributed over a space of 15,000 light years. Our own Milky Way may have looked like this about 13 billion years ago. The lensed galaxy is reminiscent of others with large redshifts in the Hubble Deep Field. The spectral lines indicate turbulent motions with velocities of 200 kilometers per second. It is conceivable that the gas is pushed back and forth by supernovae, and it may be ejected from the galaxy ultimately. This young, violent phase of galaxy formation could come to a sudden end in this way.

Astrophysics took another step toward the early universe in 1998, with the serendipitous discovery (without a gravitational lens) of a galaxy with a redshift of 5.34. For no other celestial object had such a high value ever been *measured* directly, even if the colors of a few objects in the Hubble Deep Field may point toward similarly large redshifts. Astronomers were really looking for galaxies with redshifts of around 4, but when taking a spectrum of a candidate, light from an additional object was accidentally falling through the entrance slit of the spectrograph. It showed only a single line and was visible only in infrared images of the sky. Because there was only a single, asymmetric line at 772 nanometers extending over the weak continuum, the only plausible identification was the Lyman alpha emission of hydrogen, shifted from the ultraviolet (122 nanometers) into the near infrared by the enormous redshift of 5.34.

Unfortunately, the clear jump over the magic redshift hurdle of 5 cannot be equated so easily with a true distance, because such a correlation is extremely dependent on the adopted cosmological model. For a model with a Hubble constant of 50, with no cosmological constant and at critical density, for instance, the age of the universe at which we see this galaxy would be only 820 million years, or 6 percent of the total age of the universe; assuming a lower density of 0.2, the numbers become 1.6 billion years, or 9 percent. Only limited statements about the nature of this galaxy are possible. The narrowness of the Lyman alpha line indicates that the radiation does not come from an active center but agrees much better with the characteristics of a starburst galaxy: about 10,000 hot giant stars could produce the observed light. The Keck telescope can spatially resolve the galaxy, but the more detailed investigation

A galaxy cluster as cosmic telescope. The cluster CL1358+62 was used to image a very distant galaxy — at the time of observation, it was the most distant galaxy known. At top right, the object distorted by the gravitational lensing effect is shown in detail. In the bottom right panel, it has been reconstructed mathematically. This approximates the true shape of the galaxy with a redshift of 4.92. (Franx & Illingworth and NASA) of its morphology will have to wait for the execution of planned Hubble observations. Therefore, we do not yet know whether or not this rather small object is a true protogalaxy undergoing its first episode of star formation.

The new redshift record did not last more than two months: in May 1998, a galaxy with an even higher redshift of 5.64 was discovered. This was not a serendipitous discovery but the result of a systematic program. The Keck telescopes searched for the strong Lyman alpha emission line in very distant galaxies. Narrow color filters made visible even faint galaxies, whose high redshift fell into the range defined by the filters. The enormous light-gathering power of the Keck telescopes makes such a program possible; it could not have been carried out with smaller telescopes. At the time of this writing, there are indications that this program may have discovered galaxies with a redshift of 6.5.

Cosmic Gamma Ray Bursts

About once a day, a flash of gamma rays occurs somewhere in the sky: for seconds or minutes, a point source emits extremely energetic or "hard" radiation, with even shorter wavelengths than x rays, then fades away, probably forever. Because our atmosphere completely absorbs this kind of radiation, humanity was not aware of the celestial fireworks for a long time. They were discovered as a result of the cold war, with its nuclear tests and, later, test ban treaties. Verification of the treaties was based on detection of the gamma rays released by atomic bomb tests, using an appropriate instrument on a satellite. For this purpose, in the 1960s, the United States launched the *Vela* satellites into orbit. Surprisingly, they registered gamma ray signals in the absence of any other indications, such as seismic tremors, of a secret nuclear test.

Additional detectors were placed on space probes going to other planets. It turned out that the flashes of gamma rays came from the depths of space. Because this entire subject was initially classified, the first public scientific articles about it did not appear until 1973. Now astrophysics has been pondering one of its more elusive riddles for a quarter of a century. It was not even clear at first *where* the gamma ray bursts (or GRBs) originated in the universe! Early detectors provided practically no directional information. It was not until the gamma ray detectors on several planetary space probes accurately measured the arrival times of the bursts that scientists knew which directions they were coming from. But optical observations of these areas of the sky several weeks or months after the gamma ray bursts did not reveal any obvious clues about their source. And no burst ever came from the same location twice.

Because the only data available were the development over time of the gamma ray intensity, and optical and radio observations were not able to contribute any information, there was room for many theories. The gamma ray bursts were isotropic — that is, coming from every direction with the same probability — so they could be attributed either to a very close or to a fairly distant class of objects. Speculations ranged from colliding comets in our solar system to neutron stars in the depths of the universe. A concentration toward the galactic plane, which exists for many other cosmic phenomena, or a correlation with any other type of object, simply did not exist. It was frustrating to be unable to correlate gamma ray bursts with events or objects in any other region of the spectrum.

The investigation of gamma ray bursts entered a new era with the launch of NASA's Compton Gamma Ray Observatory (GRO) on board the space shuttle *Atlantis* on April 5, 1991. This satellite circles Earth at an altitude of 450 kilometers. With its size of 4 by 9 meters and its mass of 17 metric tons, it is the "younger brother" of the Hubble Space Telescope and the second in the series of the "Great Observatories" (the other family members are AXAF, the Advanced X-Ray Astrophysics Facility, and SIRTF, the Space Infrared Telescope Facility). Compton's sensitive detectors were able to determine the approximate position of a gamma ray burst within a short period of time. These positions were communicated throughout the world, but an identification with known entities remained elusive. The positions were just not accurate enough—too many optical objects were located in the corresponding areas of the sky—to find the one that may have corresponded to the gamma ray burst. But Compton's detectors were much more sensitive than those of its predecessors, and so the catalog of gamma ray bursts increased rapidly. It became increasingly clear that not the slightest concentration existed in any direction or toward any plane. Gamma ray bursts indeed come from every direction.

By 1991, the Compton results had invalidated one of the more popular theories about gamma ray bursts. This theory attributed them to neutron stars in our own galaxy, the remnants of explosions of massive stars. Neutron stars are extremely dense spheres only a dozen kilometers in diameter, consisting of neutrons and degenerate atomic particles. Several mechanisms—infalling comets, for instance—could have produced gamma ray bursts on the surfaces of these bodies. But neutron stars should be concentrated toward the galactic plane, where massive stars are formed and die. Not only had Compton found that gamma ray bursts come from everywhere, but the intensity distribution of the bursts indicated that we seem to look beyond the volume of space from which gamma ray bursts come, since the number of faint bursts is too small.

This result eliminated neutron stars in the plane of our galaxy as possible origins, and only three general classes of objects remained as candidates for the sources of gamma ray bursts:

- Comets in our solar system. Although they are distributed spherically, it was doubtful that sufficiently energetic processes would exist to emit gamma rays.
- Neutron stars in the halo of our galaxy. While it may be possible for a large spherical halo of neutron stars around our galaxy to exist, it remained unclear how they would have gotten or developed there in sufficient numbers. Furthermore, the growing size of the catalog of positions of gamma ray bursters from Compton observations made this scenario increasingly unlikely.
- Distant galaxies in the depths of the universe. This would explain that gamma ray bursters come from every direction. In addition, the cosmological redshifts of the distant galaxies in which bursters occur would cause the smaller than expected number of faint bursts.

The Italian-Dutch x-ray satellite Beppo-SAX was launched in 1996, providing additional possibilities for gamma ray burst investigations. With this satellite, the identification of the gamma ray sources finally became possible. Immediately after a gamma ray burst, Beppo-SAX was oriented toward its approxi-

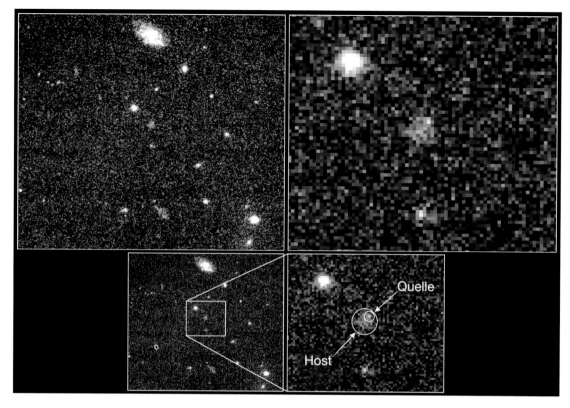

mate position to take an image with its x-ray camera. In several cases, the afterglow of the gamma ray event was visible in the x-ray domain. The positional accuracy of the x-ray camera was finally good enough to enable optical astronomers to find the needle in the haystack. Three gamma ray bursts were particularly important: GRB 970228, GRB 970508, and GRB 971214. On February 28, 1997, the gamma ray monitor of Beppo-SAX observed GRB 970228 (the number simply gives the date in the format year/month/day), followed by an image taken by the wide-field x-ray camera. Two optical observations at the William Herschel Telescope on La Palma, Canary Islands, showed a starlike object that had decreased in brightness from 21st to 23rd magnitude during the time of the burst.

Additional observations indicated that the faint source could possibly be diffuse. By mid-March, it had become a faint speck in the sky, nearly invisible to ground-based telescopes. On March 26, Hubble observed the object for the first time, identifying it as a star of magnitude 25.7 with a diffuse object in its immediate vicinity. A second HST observation took place on April 7. Confusion reigned: some scientists thought that the object had moved in the sky during the short time between the two images, others found it stationary; some observers on the ground believed they had seen a decrease in brightness of the neighboring nebula, others considered the brightness constant. Unfortunately, Hubble could not take the next image until September 5, 1997, because it was too close to the Sun in the interim. By that time,

the brightness of the point of light had diminished to magnitude 27.7, but it was still there. These observations allowed several important conclusions.

The continued visibility of the afterglow of the burst and the observed decline are consistent with theories that attribute the radiation to a "relativistic fireball" at a large distance, which expands with almost the speed of light. Such a fireball of correspondingly smaller dimension could not have occurred within our galaxy, because its expansion would have decreased rapidly by interaction with the interstellar medium, and it would have vanished rather quickly. Earlier statements about a measurable proper motion of the object, which would also have indicated a very small distance, were refuted by the third Hubble image in September. The point did not move in the sky and could therefore be at any (large) distance.

The nebulous object in the immediate vicinity of the fireball had not changed its brightness, indicating that it was not directly related to the fireball. But it could certainly be the galaxy in which the gamma

Hubble on the trail of the gamma ray burster of February 28, 1997. When Hubble took this image on September 5, it could still see the afterglow, which had declined to 0.2 percent of its maximum brightness, as well as the galaxy where the event had apparently taken place. This image may not be very impressive, but the fact that the afterglow persisted for so long is an important test for the model of gamma ray bursters. (Fruchter & Pian, and NASA)

ray burst occurred. Because the faint optical image of the gamma ray burst is located at the edge of the galaxy, the outburst apparently does not have anything to do with activities in the compact core of a galaxy. One model for the origin of the gamma ray burst involves the collision of two neutron stars; the current observations cannot prove this theory, but at least they are in good agreement with it.

In the meantime, observation of the gamma ray burst of May 8, 1997, and its afterglow provided additional information. The optical counterpart of GRB 970508 was found quickly, after Beppo-SAX had determined its position accurately. In addition, the Keck II telescope succeeded in taking a spectrum of the object. This spectrum did not show any conspicuous features at first, but more detailed analysis showed a number of absorption lines of ionized iron and magnesium in the blue-green region, with a redshift of 0.835 and a weaker absorption at 0.768. Such absorption systems with similar redshifts are usually found in the spectra of quasars. The light from the quasar crosses practically invisible galaxies on its way to us; the interstellar material of these galaxies filters these lines from the quasar's spectrum. In the case of the gamma ray burster, we can say that it has to be located at a distance corresponding to a redshift of 0.835 or more—depending on whether the burster belongs to the galaxy with the stronger absorption system or the galaxy lies in the light path just by chance. With this finding, it appears that the sources of gamma ray bursts are located at "cosmological" distances; that is, at distances of billions of light years.

The afterglow of the May burst was observed with Hubble on June 2, 1997. This time, no galaxy was visible at the location of the burst. Perhaps the burst had occurred in a very faint galaxy? Observations during the spring of 1998 finally showed a galaxy at the location of the burst; its brighter afterglow had drowned out the galaxy in the 1997 image.

The nature of the gamma ray bursts had been a riddle for a long time, but now a consistent theory based on the cosmological interpretation begins to emerge. Because the optical sources decreased in brightness fairly quickly, and because they did not exist before the bursts, their identification with the gamma ray burst of only a few minutes is virtually certain. In addition, the May burst had a counterpart in the radio domain; the radio observations provide new information about the nature of these giant explosions. The trigger mechanism (for instance, the collision of two neutron stars) for this fireball of unimaginable energy is of little consequence to its behavior during the following few weeks. The rapid transformation of energy into gamma rays, and the corresponding afterglow from the x-ray range to the optical and the radio domains, is important.

The rapid time variability in a typical gamma ray burst indicates that the source cannot be larger than about 100 kilometers in diameter initially— only about 10 times as large as a typical neutron star. These fireballs consist mainly of atomic particles erupting almost at the speed of light — a "relativistic" phenomenon, for the equations of the theory of relativity come into play. During this extremely rapid expansion, the gamma radiation emerges by some mechanism that is far from being understood in detail. The interaction of the fireball with the gas in the galaxy where the explosion occurs accounts for the longer afterglow at lower energies and also for the gradual deceleration of the fireball.

The exploding matter after the burst of May 1997 has been observed directly by two radio interferometers. The corresponding radio source "twinkled" at first like the image of a star in bad air. This was the typical "scintillation" caused by interstellar gas, an effect in radio astronomy that has been known for decades. It occurs only with extremely compact, pointlike sources, however. After about three months, the scintillation began to subside, indicating that the source had become noticeably larger. Unfortunately, not every detail is perfect: The optical light curve of the May burst does *not* fit the fireball model, at least not in its simplest form. According to the theory, it would have to reach its maximum value almost immediately, before it begins to decline; in reality, the initial increase took more than a day.

A third gamma ray burst, GRB 971214, proved that the theories are not yet accurate enough. In this case, the optical counterpart — derived from precise x-ray positions — again is associated with a faint galaxy. It turns out that this galaxy has a redshift of 3.4,

corresponding to a distance of about 12 billion light years. From this distance, the energy output of the gamma ray burst can be determined: it amounts to a staggering 10^{40} kilowatt-hours, several hundred times more than a supernova explosion, and as much energy as our own Milky Way emits over several centuries! And that includes only the gamma ray energy; it is suspected that 100 times more energy is released in the form of neutrinos and gravitational waves. A new name has been coined for these objects: hypernovae, objects that emit more energy than supernovae by several orders of magnitude. But what is the nature of a hypernova? Theoreticians tell us that the collision of two neutron stars would provide "only" 10^{38} kilowatt-hours of energy. More energy is released if the core of a massive star falls into a black hole. Is GRB 971214 such an object? Or is the assumption that the gamma ray energy is released uniformly into every direction wrong? In that case, the merger of two neutron stars would provide sufficient energy, if we only observe it from the "optimal" direction. But if we assume that the hard gamma ray burst is not isotropic, then there should be objects exhibiting only the afterglow in the x-ray domain and in the visible range. Such objects have not been found up to now — but there has been no systematic search for them, either. Observers and theoreticians will have to continue their intense efforts for the foreseeable future to explain the gamma ray burst phenomenon satisfactorily.

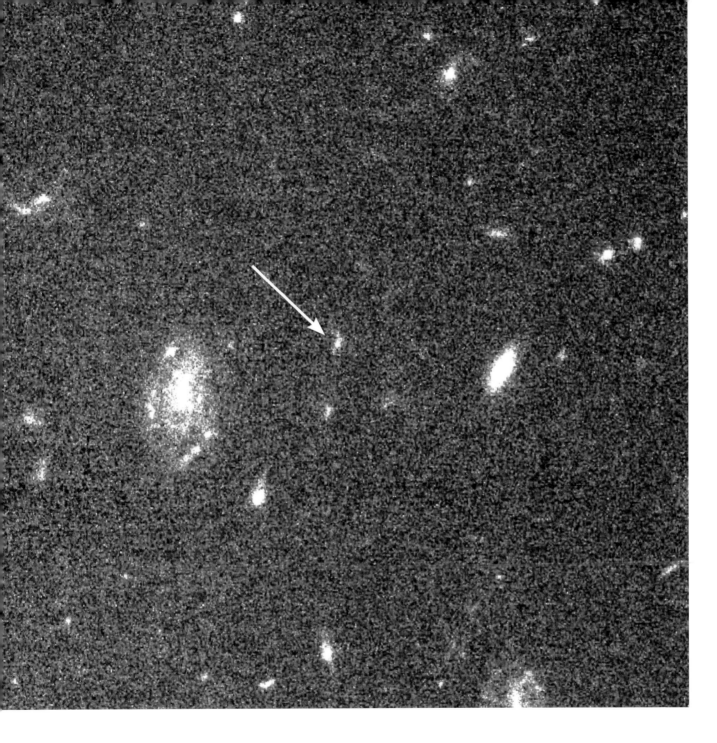

The record burst. In this inconspicuous galaxy with a redshift of 3.4, a gamma ray burst occurred in December 1997; this image was taken four months later. The afterglow has subsided and there is no longer anything to indicate the titanic explosion that took place in this object. (Kulkarni & Djorgovski and NASA)

Part 3

Stars

Stellar Nurseries

Galaxies are more than just collections of stars. The material between the stars affects the shape of a galaxy, how it behaves, and how its stars are formed. Interstellar matter consists mainly of hydrogen and helium left over from the immediate aftermath of the Big Bang, with some heavier chemical elements created in early generations of stars. The disks of spiral galaxies have an abundance of interstellar matter: in some places, it is thin and hot; in others, dense and cold. When the interstellar matter is hot *and* dense, we can see glowing nebulae, called H-II (hydrogen-II) regions. Hydrogen, the simplest atomic element, consists of one proton in the nucleus and one electron. H-II is ionized hydrogen — hydrogen that has lost its electron and acquired a positive electrical charge.

Why do these clouds glow, even though they do not generate energy themselves? Usually, H-II regions are also areas of star formation, and the newly emerged massive stars are very hot. Their ultraviolet radiation induces the hydrogen atoms to fluoresce: hydrogen atoms are ionized — that is, robbed of their electrons — and then recombine again by capturing an electron. The reddish color of these gas nebulae results from the capture of electrons by ionized hydrogen atoms. In most cases, the deeply red hydrogen alpha line is emitted.

The Orion Nebula

The H-II region closest to us is the Orion Nebula, visible in the winter sky to the naked eye as a faint cloud in the "sword" of the constellation Orion. But we only see the "hot" tip of the iceberg: the Orion Nebula is still mainly a huge, cold, dusty molecular cloud, largely impenetrable by visible light. In its interior, gas clumps coalesce into new stars, emitting infrared radiation in the process. Such stellar nurseries cannot be seen by optical telescopes, and until the second servicing mission, they were inaccessible even to Hubble.

But for several years now, the Hubble Space Telescope has provided images of the Orion Nebula with previously unobtainable resolution. These images show young stars in front of a glowing background nebula, surrounded by protoplanetary disks (called proplyds). Let us take a look at a mosaic of the Orion Nebula composed of 15 individual images from the Hubble Space Telescope. It covers an area of 30 square arc minutes; given a distance of 1500 light years, one side of the image corresponds to 2.5 light years. The four hottest, most massive stars of the Orion Nebula form the "trapezium." They are largely responsible for the excitation of the nebula and therefore for its glow. In addition to the trapezium stars, there are about 700 other young stars at various stages of evolution — together, they form a truly remarkable cluster. The stars in this cluster are packed 10,000 times more densely than the stars in the vicinity of the Sun. Several stars developed high-velocity gas jets that caused shock fronts in the thin gas of the nebula; they can be seen in the image as

A color panorama of the central region of the Orion Nebula, composed from 15 individual WFPC2 images. We see young stars, and the effects of their formation are visible as shock fronts and focused streams of matter, giving the nebula its extremely complex shape. (O'Dell and NASA)

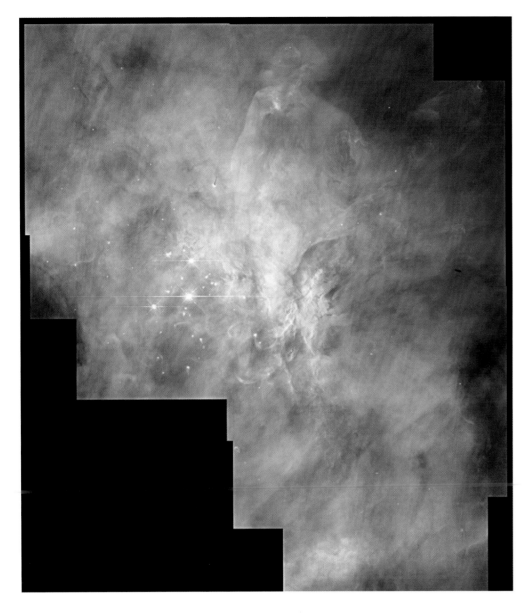

The inner part of the Orion Nebula, with the trapezium stars and other young stars, shows the complex gas clouds from which these stars emerged. (Johnstone & Bertoldi and NASA)

thin, curved structures.

The Orion Nebula mosaic contains about 500 stars, 153 glowing protoplanetary disks, and a few *dark* proplyds, silhouetted against the brighter gas nebula. Although the proplyds consist of 99 percent gas and only 1 percent dust, the small amount of dust is enough to make them totally opaque. The

stars in the interior of the proplyds have masses of 30 to 150 percent of the Sun's mass. The proplyds are embryonic solar systems, and planets may form there in the future. Proplyds appear to be the rule rather than the exception: almost every third star in the trapezium cluster is surrounded by a visible disk, and additional stars show at least some indication of dust in the infrared. In total, half the stars in the Orion Nebula may be accompanied by a surrounding disk—which leads to the conjecture that the formation of planets around stars is a common occurrence in the universe. There may be factors, however, that inhibit planetary formation. Proplyds close to the trapezium stars lose part of their gas and dust as a result of the radiation pressure from these stars. Model calculations show that the gas and dust are blown away from the star within a million years, whereas the formation of planets takes about 10 times longer.

The new NICMOS infrared camera is the first to penetrate the depths of the dust-enshrouded Orion molecular cloud. A comparison of optical and infrared images shows a prominent starlike object in the infrared, which is invisible in the optical range. This is the Becklin-Neugebauer object, one of the first infrared sources observed in the Orion Nebula. In contrast, the brightest stars in the WFPC2 image, taken in visible light, are the four familiar trapezium stars. The infrared image shows several extended, previously unknown structures. A dark, crescent-shaped area above (north of) the Becklin-Neugebauer object is possibly a clump of matter, induced to

glow by some radiation source or by a stream of matter. Two additional conspicuous, bright arcs below (south of) the Becklin-Neugebauer object are radiating interstellar dust, possibly associated with streams of matter from young stars. An attentive observer notices three pairs of close binary stars in the infrared image. The projected distance between the components of a pair corresponds only to twice the distance between the Sun and Pluto, the outermost known planet in the solar system.

The Eagle Nebula

Hubble took one of its best known and most fascinating images in the central part of the Eagle Nebula (M16) in the constellation Serpens (the snake). This object is about 7000 light years away. The eagle's nest contains many "EGGs"; this term has nothing to do with real eggs, of course, but stands for Evaporating Gaseous Globules—maybe astronomers do have a sense of humor. The EGGs are located at the ends of fingerlike structures that are rooted in giant columns of cold gas. These columns, called "elephant trunks," emanate from the surface of a huge cloud of cold hydrogen. Within the columns, which are several light years in diameter, the cool gas is dense enough to collapse under its own gravity, forming young stars. These accumulate more and

Two versions of Hubble's close-up of the trapezium stars. In addition to the four bright stars and several other young stars, a number of proplyds are visible. The strong radiation from the stars has distorted some of them. (Johnstone & Bertoldi and NASA)

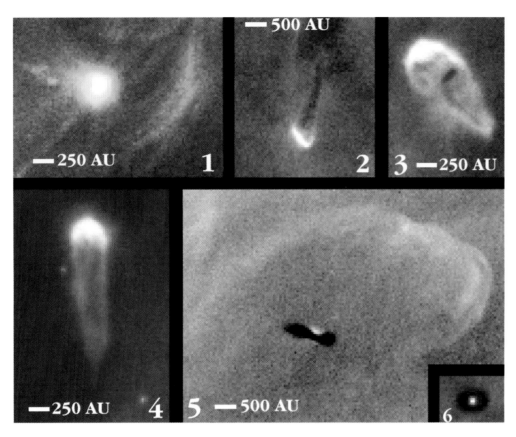

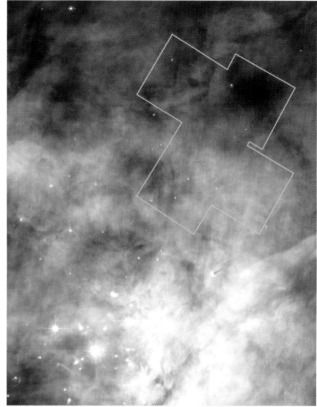

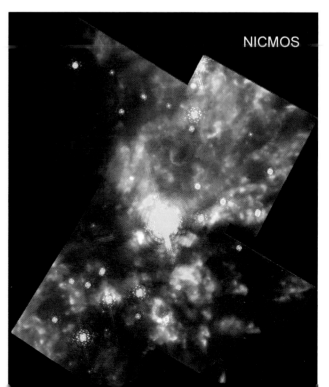

A gallery of circumstellar disks from the Orion Nebula. The first four show the effects of the central stars on the disks; the remaining two appear as silhouettes in front of the bright nebula. (Johnstone & Bertoldi and NASA)

The heart of the Orion Molecular Cloud (OMC-1) unveiled by Hubble's NICMOS infrared camera. The WFPC2 image (top) shows little detail, but the infrared image (bottom) provides a view of a chaotic star-forming region with a massive young star. This Becklin-Neugebauer object has been known for some time—it is one of the brightest sources in the infrared sky—but it is invisible in visual light. (Thompson et al. and NASA) more mass from interstellar matter in their vicinity. As the first massive young star ignites, it emits a flood of ultraviolet radiation, heating the gas on the surface of the columns and evaporating it into interstellar space. The Hubble image shows this gas, evaporating under the influence of intense light, as ghostly rays flowing away from the columns.

But the gas does not evaporate equally at all locations; the EGGs are denser and provide more resistance to the flood of light. Several EGGs appear as small humps on the surface of the columns; others are more exposed and look like fingers emanating from the larger cloud. These fingers consist of gas that is shielded from evaporation by the shadow of the EGGs. Some fingers have detached themselves completely from the column and resemble teardrops. And in some cases, the stars in the EGGs' interiors can be seen.

This impressive image, giving a strong sense of depth, became *the* Hubble image in November 1995, even if it provided only limited new insights: The "elephant trunks" are so large that their picture can be taken even with small amateur telescopes. But the Hubble image did sharpen the discussion about the inner workings of the gas globules. Do stellar embryos grow in every EGG? Or is the density insufficient to lead to gravitational collapse, so that star formation is suspended in the Eagle Nebula at this time?

Only infrared astronomy can provide the answer, because it can look into these potentially star-forming clouds. First analyses appeared to support the interpretation of EGGs as homes for stellar embryos (also called Young Stellar Objects, or YSOs). The large team of authors of the first Hubble-based publication on the Eagle Nebula speculated that a good fraction of EGGs contained YSOs: some of those had apparently been seen in earlier infrared images; even in visible light, stars could be seen in some of the EGGs. This team even surmised that the disklike proplyds in the Orion Nebula were really EGGs. This led to a controversy with the proplyd scientists: they saw the Orion Nebula as the birthplace of many planetary systems and objected to the radical reinterpretation of these objects as globules. They were able to provide several arguments for the disk-shaped structure of the proplyds; the corresponding images were quite obvious in that regard. The premise that many of the EGGs contained young stars was also questioned: in the literature, the Eagle Nebula was not considered to be a place of currently active star formation. Only new infrared observations with better resolution could help to solve this controversy. Because Hubble had not yet received its NICMOS camera, ground-based telescopes and instruments had to be used to investigate the interiors of individual EGGs. The 3.6-meter telescope of the European Southern Observatory in La Silla, Chile, achieved near-infrared images with a resolution of 0.13 arc second, by applying adaptive optics and computer image enhancement.

The data from May 1996 were quite clear. Except for one case in which a young stellar object indeed appears to be located in an EGG, all the other EGGs are empty — no stars are forming there at the moment. If embryonic stars existed there, they would have attracted attention as conspicuous point sources at a wavelength of 2 microns. The point sources that do appear in the images come from background objects and have no connection to the EGGs. The previously assumed positions of EGGs and stars were based on infrared images of inferior resolution and had now been superseded. These EGGs are now called "preprotostellar regions" to indicate that they are indeed dense regions of the nebula, contracting under their own gravity, where stellar formation could begin at a later time. On the other hand, stars may have begun to form already but are still too small and too cool to be observed in the near-infrared domain.

The Eagle Nebula has remained an important target for infrared astronomy, both from the ground and from space. NICMOS images from Hubble have not yet been taken, but new data are available from ISOCAM, the camera on board the European Infrared Space Observatory. These images cover the wavelength regions from 9 to 12 and from 12 to 18 microns. Again, they provided no indication of young stellar objects inside the EGGs, except for one case. Indeed, the star formation rate was interestingly low: the same process that reduces the molecular gas in the nebula and leaves behind only the dramatic elephant trunks must also inhibit the formation of stars. It is still possible, however, that we are witnessing only the onset or earliest stages of star

development. Recent data also show that EGGs are not the same phenomena as the proplyds in the Orion Nebula. In almost all of the proplyds, stars have been detected in the infrared.

The Cone Nebula and the Lagoon Nebula

Hubble images taken of the Cone Nebula with the new NICMOS infrared camera are less controversial. Although nothing at all is visible in the optical wavelength range, the near infrared shows a massive, bright star (NGC 2264 IRS) that was already known and six previously unknown "baby" stars. These baby stars are located in the immediate vicinity of the bright star, at distances of only 0.04 to 0.08 light year (400 to 800 billion kilometers). NGC 2264 IRS is 2000 times more luminous than our Sun and emits a strong stellar wind; it continuously loses mass into the surrounding space. This stellar wind compressed the neighboring interstellar matter and triggered the formation of the six much smaller, more Sun-like stars. This image also reveals a technological triumph: the diffraction rings around the bright and faint stars told the NICMOS team that the camera's optical system works perfectly.

Another stellar nursery lies in the Lagoon Nebula (M8), an object that is just visible to the naked eye at a distance of 5000 light years in the constellation Sagittarius (the archer). The central star, Herschel 36, induces parts of the gas to glow with its energetic radiation. The bright gas cloud is also called the

The "elephant trunks" of the Eagle Nebula (M16) and the EGGs (Evaporating Gaseous Globules) at their ends provide dramatic evidence of the gradual destruction of a molecular cloud under the bombardment of radiation from stars that developed there. The initial suspicion that the EGGs (the concentrations emanating from the gas columns, shown enlarged on the right) contain emerging new stars has not been confirmed. (Hester & Scowen and NASA)

Hourglass Nebula, because of its characteristic shape. Hubble's view of this scene reveals a multitude of small structures in the dense interstellar gas: small dark clouds (called Bok globules), arched shock fronts around stars, fragments of ionized gas, gas rings, and rays and knots of gas condensation. And there is something surprisingly reminiscent of a tornado! In the core of the nebula, a pair of vortexes about half a light year in diameter can be seen in the gas that is ionized by hot stars. Herschel 36 could indeed have caused a phenomenon similar to tornadoes on Earth. The large temperature difference between the hot stellar surface and the cold interior of the dense gas clouds, together with the radiation pressure, may have caused a strong horizontal shear that could have twisted the gas clouds into the suggestive shape of a tornado.

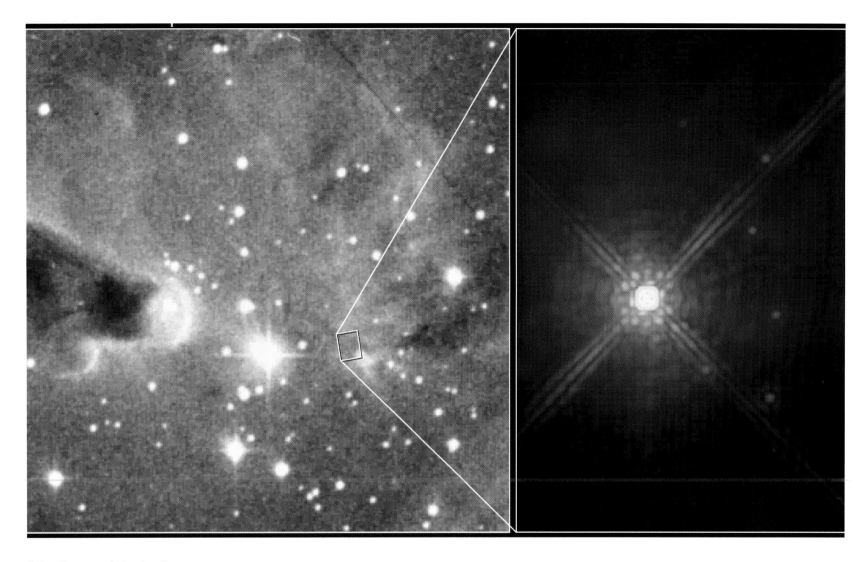

A family portrait in the Cone
Nebula, taken by the NICMOS
camera. Six young stars
about the size of the Sun
are visible around the more
massive star NGC 2264 IRS,
which apparently triggered
their formation. (Thompson
et al. and NASA)

"Tornadoes" in the Lagoon Nebula? The gas flows in the center of this nebula are reminiscent of terrestrial tornadoes — and the underlying physics may be quite similar. This investigation was based on Hubble data taken for a completely different program but later retrieved from the Hubble Data Archive for this purpose. (Caulet and NASA)

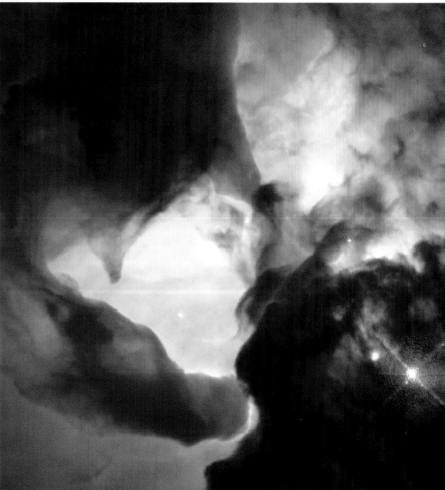

Cosmic silhouettes, dust in galaxies made visible. Since 1990, scientists have hotly debated whether dust in the spiral arms of galaxies is opaque or only partly absorbs the light from behind. The answer was provided by this 1998 Hubble image of spiral arms that are coincidentally located in front of other galaxies. The dust is concentrated in clumps, and even the densest regions still allow 20 percent of the light to shine through. (Keel & White and NASA)

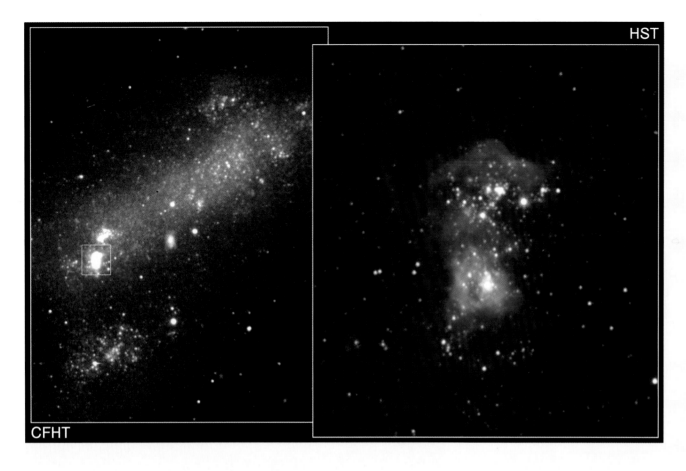

Active star formation in a faint galaxy. The irregular galaxy NGC 2366 looks rather inconspicuous, except for the bright star-forming region NGC 2363. The WFPC2 image shows two dense clusters of massive stars of different ages. The one on top is 4 to 5 million years old and has lost most of its gas; the bottom cluster, less than 2 million years old, still remains within the nebula from which the stars formed. (Drissen et al. and NASA)

A Distant View into Star-Forming Regions in Other Galaxies

Hubble Space Telescope investigations of star-forming regions have ranged far beyond the relatively small ones near our Sun, such as the Orion Nebula. There are particularly active and extended star-forming regions in "irregular" galaxies like the Magellanic Clouds. These galaxies, which have relatively low total mass, are only now undergoing their maximum rate of star formation. By contrast, our Milky Way and other large spiral galaxies went through this phase of their evolution in the distant past.

NGC 2366 is such an irregular galaxy. Its large star-forming region received a separate number in the catalog of galaxies—NCG 2363—because of its conspicuous appearance. This galaxy is 10 million light years away from us. NGC 2363 is comparable in size to the 30 Doradus Nebula in the Large Magellanic Cloud, a magnificent star-forming region that dominates the appearance of this neighbor of our galaxy.

The brightest star in Hubble's image of NGC 2363 is a massive luminous blue variable of 30 to 60 solar masses that is currently undergoing an active phase. Only Hubble's superb resolution can clearly distinguish this star from the others nearby. A comparison with archival images shows that the star increased in brightness by a factor of 40 over three years. Such stars and their outbursts are relatively rare—counterparts in our own galaxy include P Cygni, which had an outburst around the year 1600, and Eta Carinae,

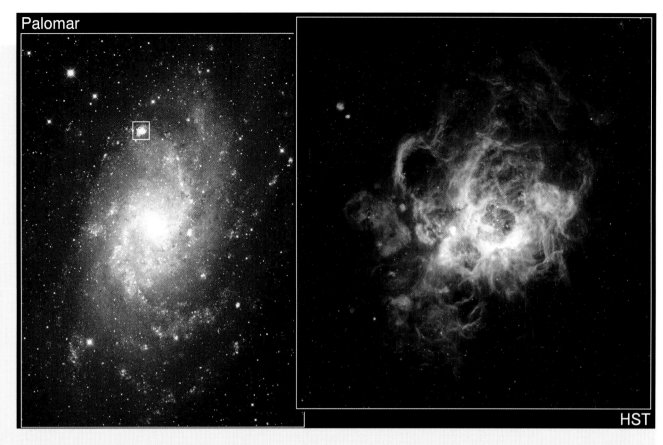

Palomar

HST

Stars are being born in the gigantic nebula NGC 604 in the spiral galaxy M33. More than 200 young stars with 15 to 60 solar masses have been discovered. Their radiation illuminates the nebula and also affects its shape. (Yang and NASA)

which was active from 1837 to 1860. The HST image also shows two dense star clusters composed of massive stars at different stages of their evolution. Stellar winds and shock waves have removed the gas from the older (top) cluster, which is 4 to 5 million years old. The younger and brighter cluster (center and bottom) has an age of less than 2 million years and is still embedded in the gas and dust clouds from which it emerged.

It is interesting to speculate about why such active star formation occurs at the end of the central bar in irregular galaxies. Apparently, the motions in the bar concentrate the gas at its ends, leading to conditions conducive to the development of stars. The efficiency of this mechanism is surprising: the star-forming region

is 10 times bigger and 10 times brighter than the most prominent one in our own, much larger galaxy!

Another giant star-forming region, NGC 604, is located in the spiral galaxy M33, at a distance of 2.7 million light years in the constellation Triangulus (the triangle). M33 is a member of the Local Group of galaxies, to which our Milky Way also belongs. Although it is normal that new stars form in the spiral arms, this area is particularly large: it has a diameter of 1500 light years. The center of NGC 603 contains more than 200 hot stars with masses between 15 and 60 times that of the Sun. Their energetic radiation induces the gas in the star-forming region to glow.

Giant Stars

Betelgeuse

"All stars appear only as point sources even in the largest telescopes." This "wisdom" of older astronomy textbooks has been inaccurate for quite some time. Even at the beginning of the twentieth century, clever methods existed to determine the angular diameter of the largest stars in the sky. Since the 1970s, attempts have been made to obtain resolved images of stars using sophisticated optical and mathematical methods. But it was still exciting when Hubble and its Faint Object Camera provided the first *direct* ultraviolet images of another star. The target was Betelgeuse, or Alpha Orionis, the bright red supergiant star marking the shoulder of the winter constellation Orion, the hunter. Betelgeuse is so huge that, if it replaced the Sun at the center of our solar system, its outer atmosphere would extend past the orbit of Jupiter. The Hubble images show that the shorter the wavelength range used for the observations, the larger the star's diameter appears to be. This phenomenon is caused by the extended, thin atmosphere of this red giant: the longer the wavelength, the more deeply we probe this atmosphere. In green light, Betelgeuse appears to have a diameter of 0.05 arc second, whereas in the ultraviolet, it appears to have a diameter of 0.11 arc second. The Faint Object Camera has a theoretical resolution of 0.015 arc second since the installation of COSTAR.

Thus, Hubble can resolve Betelgeuse easily, and interesting details immediately began to appear. Right in the middle of the atmosphere, a mysterious bright spot, 2000°C hotter than its surroundings, was visible on the first image. Additional observations confirmed the existence of one or more immense hot spots on this giant star. They are expected to correspond to huge convection cells, such as exist on a much smaller scale in the Sun: hot gases well up from below, lose energy, and sink down again. This Hubble observation helped to explain another of Betelgeuse's enigmas: radio astronomers observed the existence of cool bubbles in the extended atmosphere of the star. Apparently, the strong convection ejects entire pockets of cooler gas high into the atmosphere—and the hot spots in the Hubble images may just be the locations from which they were ejected. The strong stellar wind that emanates from giant stars such as Betelgeuse may be driven by these convection processes.

The Pistol Star

One of the first observations with the new NICMOS camera was of a star at a distance of about 25,000 light years, close to the center of our galaxy and therefore shrouded by large amounts of interstellar dust. It is called "Pistol Star," receiving its name from the pistol-shaped nebula surrounding it. This star is 10 million times more luminous than the Sun: in one second, it emits as much energy as the Sun does in two months! The Pistol Star would be

The first direct image of the surface of another star. The Faint Object Camera successfully observed Alpha Orionis, also known as Betelgeuse, the bright red star in Orion. Because is it a red giant, it is a particularly large star, which allowed FOC to resolve details of its extended atmosphere and discover an extended bright spot. (Dupree & Gilliland and NASA)

Translation of German terms (English in parentheses):
Größe des Sterns   (Size of Star)
Größe der Erdumlaufbahn (Size of Earth's Orbit)
Größe der Jupiterbahn   (Size of Jupiter's Orbit)

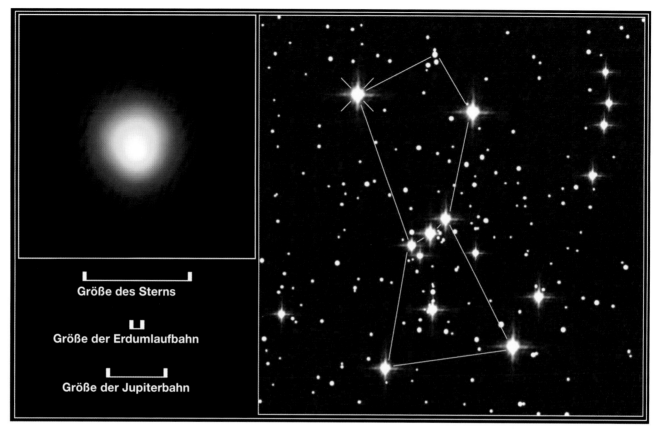

Größe des Sterns

Größe der Erdumlaufbahn

Größe der Jupiterbahn

visible to the naked eye as a fourth magnitude star, if it were not for interstellar dust clouds of tiny particles between us and the center of the Milky Way that absorb the star's light. Even the most powerful telescopes cannot see the Pistol Star in visible wavelengths. In the infrared, about 90 percent of the radiation is still absorbed by dust, but the remaining light provides an interesting picture.

A star with such luminosity must be extremely massive; otherwise, it would be torn apart by its own radiation pressure. Estimating the mass from the luminosity yields a value of 150 solar masses. Despite this colossal supply of nuclear fuel, the star will have exhausted its resources over a lifetime of only millions of years, compared with the Sun's lifetime of billions of years. Massive stars are so luminous that they consume their fuel at an outrageous rate, burn-

ing out quickly and often creating dramatic events. As these stars evolve, they can eject substantial portions of their atmospheres—in the case of the Pistol Star, producing the surrounding nebula and an extreme stellar wind that is 10 billion times stronger than our Sun's. At some point in time, the nuclear fuel in the center of the star will be used up, and the star's core will collapse, ejecting the outer layers into space as a supernova. Burning at such a dramatic rate, this climactic death of the Pistol Star may occur in about one to three million years.

But this star has already undergone violent phases in the past, as shown by the surrounding gas nebula. The Pistol Nebula, named for the characteristic shape that also gave the star its designation, has an extent of about four light years, comparable to the distance of the closest star from the Sun. Astronomers believe

The Pistol Star, one of the brightest stars of our galaxy, in front of its nebula, as seen by NICMOS. Because it is located close to the galactic center, its bright light is strongly absorbed by dust in the visible wavelength range. (Figer and NASA)

that the Pistol Nebula was created by eruptions in the outer layers of the star which ejected up to ten solar masses of material in giant outbursts about 4000 to 6000 years ago.

Theoreticians will have a difficult time explaining the origin of such a massive star. And theoreticians and observers will have to cooperate in the investigation of its long-term stability. Current estimates of the star's luminosity, and therefore of its mass, are still uncertain, however — and it is still conceivable that we are seeing several extremely concentrated smaller stars. But one way or the other, we are undoubtedly seeing a true beacon in space, albeit through an incomprehensibly dense nebula.

Eta Carinae

Another superstar in our galaxy is Eta Carinae in the constellation Carina, close to the Southern Cross; the star and its impressive gas nebula can be seen only from the southern hemisphere. Edmund Halley had cataloged this object as a star of fourth magnitude in 1677. In 1730, however, it reached second magnitude, then faded and became brighter again, reaching its maximum in April 1843, when it became the second brightest star in the sky after Sirius. The energy output of this mysterious star is enormous: it emits as much light as about 5 million Suns. This increase in brightness also corresponded with the ejection of matter that condensed to dust. An instability had probably occurred in the outer layers of this luminous blue variable; such stars are barely able to withstand their own extreme radiation pressure. The resulting dust attenuated the light of the star for decades, and its visual brightness sank to eighth magnitude, rebounding slightly over the last few years.

Despite Eta Carinae's distance of about 8000 light years, modern image processing techniques extracted an extremely clear picture from the Hubble observations, showing details on the scale of our solar system. The Hubble image provides information about the geometry of the ejected matter. Two polar clouds and a large, thin equatorial disk expand into space at a velocity of 650 kilometers per second. The new images show ultraviolet light coming mainly from the equatorial region, because it contains less absorbing dust. In contrast, the polar clouds include more dust and therefore appear reddish. Dust lobes, minute condensations, and peculiar radial streaks appear with uncommon clarity. Eta Carinae is expected to end its life as a giant star in less than 100,000 years; it will become a supernova, leaving a black hole or a neutron star as remnant.

Astronomers also made visible the furious expansion of the huge, billowing pair of gas and dust clouds. They aligned and subtracted two images of Eta Carinae taken 17 months apart, in April 1994 and in September 1995. The resulting difference image is remarkable because most celestial objects barely change noticeably over a span of many years. Eta Carinae is a dramatic exception because it underwent such a titanic explosion 150 years ago. Scientists can now track the motions of hundreds of small-scale structures in the lobes and characterize their evolution. This may give clues as to how the lobes formed in the first place, and shed light on the bipolar phenomenon in general.

A Swedish scientist also discovered a laser mechanism at work in a gas bubble in the immediate vicinity of Eta Carinae. Unusual spectral emission lines from this bubble could be traced back to the element iron, but the relative strengths of these lines did not conform to expectations. A phenomenon called stimulated emission is the only plausible explanation: an effect exactly equivalent to that in a laser source enhances the strongest lines even more. From the gas bubble, in which the laser process occurs, concentrated rays of light shoot out in various directions. One of these laser rays was pointed exactly toward Earth.

The enigma of Eta Carinae also has a temporal component: two pronounced periodicities characterize the behavior of this star. Its spectrum changes dramatically every five and a half years, and a maximum appears in its x-ray emission every 85 days. The x-ray intensity increased continuously after 1996, only to decline dramatically at the end of 1997. Models that explain all these phenomena are still rudimentary, but it appears that Eta Carinae is a binary system with an orbital period of 5.5 years. Around the end of 1997, the two objects reached their closest approach. At the beginning of the 1980s and again in the middle of 1992, spurious changes were observed in the spectrum of Eta Carinae: the emission lines suddenly became weaker or disappeared for weeks, while the radio and x-ray signals underwent strong variations at the same time. Another one of these "crises" had been predicted for the end of 1997—and indeed the highly excited spectral lines disappeared within a week of the time forecast. The 5.5-year cycle had been established, and the clocklike behavior of this superbright star can only be explained if it is a binary system.

Eta Carinae is therefore the most massive known

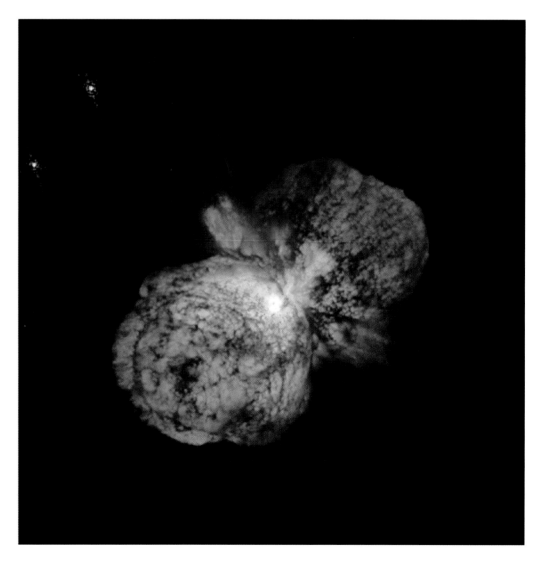

The sharpest image of the Eta Carinae Nebula. The high resolution of Hubble's camera and complex image processing techniques provide fine details of this strange nebula, which resulted from the eruption of a star 150 years ago and has expanded ever since. The continuing increase in size can be made visible by subtracting two images, taken 17 months apart (right). Matter closer to the star expands faster than material farther out. (Morse & Davidson and NASA)

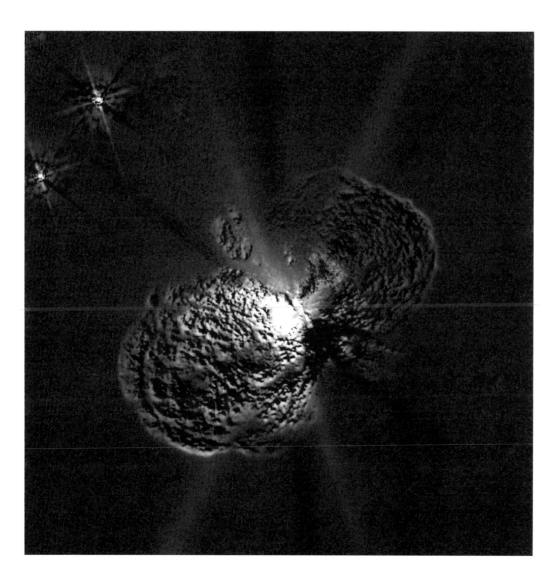

binary system; each of the two components weighs in at about 70 solar masses! The spectral changes on a 5.5-year cycle can now be explained. The closer the two stars approach each other in their orbit, the more dramatically their stellar winds—with velocities of 500 and 1000 kilometers per second—collide. The collision increases the x-ray luminosity, and the emission lines get absorbed in the increasingly dense gas. But what causes the prominent x-ray flares every 85 days? And what caused the sudden decrease in the x-ray emission at the end of 1997? The decrease may be a result of dust around one of the two stars absorbing the x-rays at the time of their closest approach. The x-ray flares, however, have no counterpart in the optical part of the spectrum and remain mysterious. Additional Hubble observations in the future may help to solve this puzzle.

Globular Clusters, White Dwarfs, and Blue Stragglers

Blue Stragglers

One of the fundamental concepts in astronomy is that the stars of a given cluster formed at about the same time. This can be seen easily in a Hertzsprung-Russell diagram, where the colors of the stars are plotted against their brightnesses. The diagram is not evenly populated but shows several preferred areas; the main sequence, corresponding to the adulthood of a star, stretches from bright blue stars to faint red stars. Depending on the age of the cluster, a smaller or larger number of the bright blue stars may be missing from the main sequence. These more massive stars develop more quickly than less massive stars, and they migrate from the area of the bright blue stars in the diagram to that of the bright red giants. Finally, the outer layers of a red giant are lost, and a white dwarf emerges. White dwarfs are difficult to observe in globular clusters, because the clusters are located so far from our Sun and white dwarfs emit only small amounts of light. Therefore, it makes sense to search for these stars in the closest globular cluster, M4. In M4, eight white dwarfs were discovered in a small field with a diameter of only 0.63 light year, and 75 were found in a somewhat larger area. Extrapolation shows that the entire globular cluster contains about 40,000 white dwarfs.

The Hertzsprung-Russell diagram can be used to determine the age of a globular cluster, given our knowledge of how a star evolves. In general, globular clusters are very old structures, even though some

of them may be formed in galactic collisions. On the other hand, open star clusters in the disk of our galaxy come in all ages. More massive stars age more quickly; they leave the main sequence, which is still populated by normal, less massive stars, and become red giants. In the globular cluster 47 Tucanae, for instance, all stars with masses of more than 0.9 solar masse have evolved to the red giant stage. Only the lowest part of the main sequence is still populated by yellow and red dwarf stars; the upper part, the region of blue stars, is practically empty. If we examine the area of the blue stars in the Hertzsprung-Russell diagram of a globular cluster more closely, we can

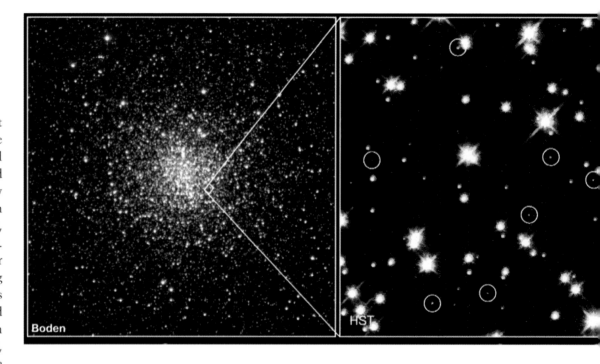

White dwarfs in the globular cluster M4, shown in a ground-based image on the left. About 40,000 of the 100,000 stars in this cluster are white dwarfs. Hubble found 75 of them in a small area; seven are marked in the HST image on the right. (Bolte & Richter and NASA)

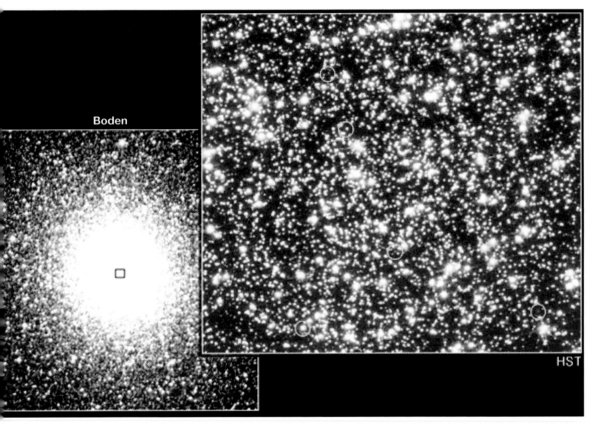

HST

"Blue Stragglers" in the globular cluster 47 Tucanae, shown in a ground-based image on the left. The Hubble Space Telescope identified several of these objects in the crowded center of the cluster (right). Blue stragglers probably developed from the merger of two stars. (Saffer & Zurek and NASA)

see a number of leftover objects—the so-called blue stragglers.

The blue stragglers are still located on the main sequence, and because they are fairly massive stars, they must have formed much more recently than the cluster. How can this be explained? Everything points to the principle that all members of a cluster developed at the same time—say, some 15 billion years ago. How can there be younger blue stragglers? Such a star near the center of 47 Tucanae was examined in detail with Hubble's Faint Object Spectrograph. The spectrum of a star indicates temperature, rotation rate, surface gravity, and—

given brightness and distance—diameter and mass. The blue straggler in the heart of 47 Tucanae is a star of 1.7 solar masses, rotating relatively quickly on its axis. All the other stars with such a mass long ago moved through the red giant phase and then developed into white dwarfs, but this latecomer is still sitting on the main sequence.

By all indications, blue stragglers result from the recent fusion of two lower mass dwarf stars that did not develop past the main sequence. Thus, blue stragglers are not born later than all the other stars of a cluster. Like the other stars, they too originated some 15 billion years ago—only not as a single star, but as two components. Two dwarf stars of low mass joined together only recently (several million years ago) to form a new object, the blue straggler. Most likely, this fusion did not result from a violent collision of two stars with completely different paths but from the gradual joining of two objects that had orbited each other closely.

Collapsing Cores

Our interest in globular clusters is not limited to the peculiar fates of single stars within them. Their development as an entirety over billions of years is also fascinating. The Hubble Space Telescope is an ideal instrument to investigate these star-packed objects. In the M15 globular cluster, more than 30,000 stars are visible in an image of an area with a diameter of only 28 light years. And the concentration of stars in-

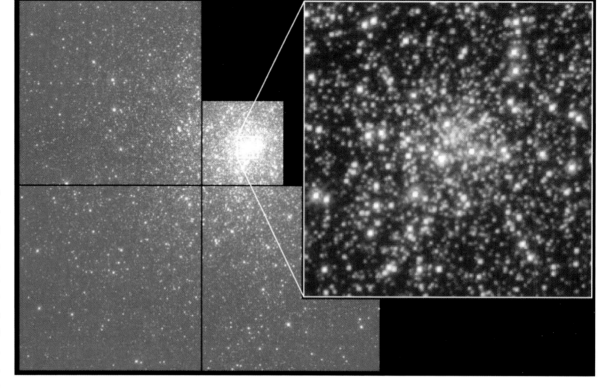

creases even more toward the center of the cluster. There are two possible explanations for this increase of star concentration near the center. Perhaps a black hole of several thousand solar masses in the center of the cluster forces the higher concentration of stars by its gravitational pull. Alternatively, the core of M15 may have collapsed. Model calculations have predicted such a phenomenon for some time. Within a few million years, an instant in the several-billion-year life of the cluster, a major instability may have developed.

Such a catastrophic event occurs when stars near the cluster center lose some of their energy of motion as a result of the gravitational forces they exert on one another. After a few billion years, some stars become too "lethargic" to withstand the gravitational forces of their neighbors and are drawn more and more quickly toward the center of the cluster. Other processes (binary stars, for instance)

somehow prevent a complete collapse of the core, so that a new, more concentrated—but stable—configuration is reached. About every fifth globular cluster is suspected of having undergone such a core collapse. Since 1991, Hubble has provided detailed images of several dozen globular clusters, but in none of them is the concentration toward the center as pronounced as it is in M15. Thus, M15 may provide the best evidence to date for a core collapse.

The core of the extremely dense globular cluster M15. Hubble measurements provided the number of stars per unit volume as a function of distance from the cluster center, confirming the suspicion that a core collapse had taken place. (Guhathakurta et al. and NASA)

Still Going Strong: Supernova 1987A

The expansion of Supernova 1987A. As the intricate ring system glows, the cloud from the explosion in the center expands so rapidly that Hubble can measure its growth directly. Remarkable changes in the various shapes have also been observed since 1995. (Pun & Kirshner and NASA)

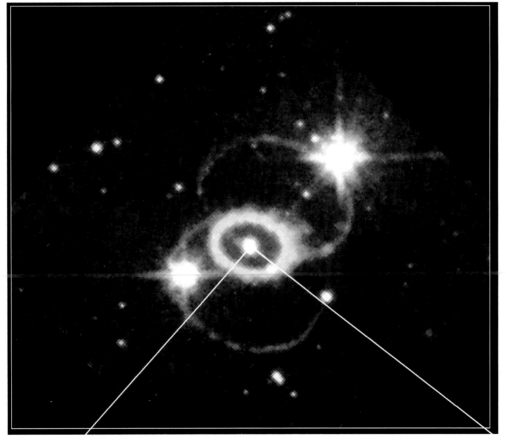

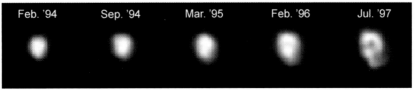

| Feb. '94 | Sep. '94 | Mar. '95 | Feb. '96 | Jul. '97 |

The explosion of a blue supergiant in the Large Magellanic Cloud on February 21, 1987, was the first supernova that modern astrophysics was able to observe in one of the close companions of our own galaxy. This point of light at the edge of a small galaxy reached an apparent brightness of second magnitude within a few weeks and then faded over the years into such a faint object that it can only be studied with HST. The light from the supernova took 167,000 years to reach us from the Large Magellanic Cloud. A massive star had exhausted its nuclear fuel, its inner core collapsed, and the energy released in this collapse ejected the star's outer layers at speeds of up to 15,000 kilometers per second. In the early phases of the explosion, large quantities of energetic radiation were released, inducing the gas in the vicinity of the supernova to glow, just as if somebody had switched on a gigantic fluorescent light.

This glowing gas exhibits a strange structure: a relatively dense, small ring is surrounded by two thin, extended rings. The inner ring is comparatively easy to explain—the progenitor of the supernova ejected this material into space when the star was still a red supergiant. It is more difficult to understand how it obtained its toroidal shape. The strong stellar wind of the star, which had now become a blue supergiant, may have pushed the older gas into its current configuration. The exact origin of the two outer rings is still unclear. Analysis of Hubble images showed that a weak glow also comes from the area inside the rings. Perhaps the rings represent only the outer edge of an hourglass-shaped shell. Chemical analysis indicates that the material of the outer rings is less "processed" (in the nuclear reactions in the core of the star) than that of the inner ring. Therefore, the outer material must have originated from the star when it was still a red giant.

More than 10 years later, the Hubble Space Telescope still provides interesting details about supernova 1987A in the Magellanic Cloud. Hubble not only can resolve the expanding shell of matter from the exploding star but also can determine its shape: it is not spherical, but consists of two bubbles in the shape of a dumbbell. The outer edges of the bubbles expand at a speed of 2700 kilometers per second and have now reached a diameter of a tenth of a light year. The plane of the "waist" of the dumbbell-shaped shell coincides with the location of the inner ring around the supernova. Apparently, a large amount of matter—which was compressed into the inner ring—had been concentrated in this plane and now impedes the expansion of the exploding shell. The spherical asymmetry of the explosion may also be related to rapid rotation of the former star or to the presence of an undiscovered companion.

The visible exploding cloud does not contain the entire mass that has been expanding into space since 1987. There are indications that some of the matter has traveled much farther. Contact between the rapidly expanding matter and gas in that area has already occurred, as indicated by radio signals and by x-ray emissions observed by ROSAT since the early 1990s. During the days following the explosion, the most rapidly expanding gas was observable in the various spectra but then was absent for the next 10 years because of its low density. Far-ultraviolet STIS spectra demonstrate that the interaction of this hydrogen, moving at speeds of 15,000 kilometers per second, with gas still inside the inner ring has now begun. The material ejected from the supernova collides with this gas and heats it to temperatures of about 50,000 K. Characteristic Lyman alpha emission from hydrogen is visible in an extended zone within the inner ring, together with nitrogen emission. In addition, the STIS spectra contain several emission lines from the hot ring itself.

The next phase of the collision has already started. One small part of the inner ring has increased in brightness since 1997 and now forms a conspicuous bright spot. In this area, a small inward extension

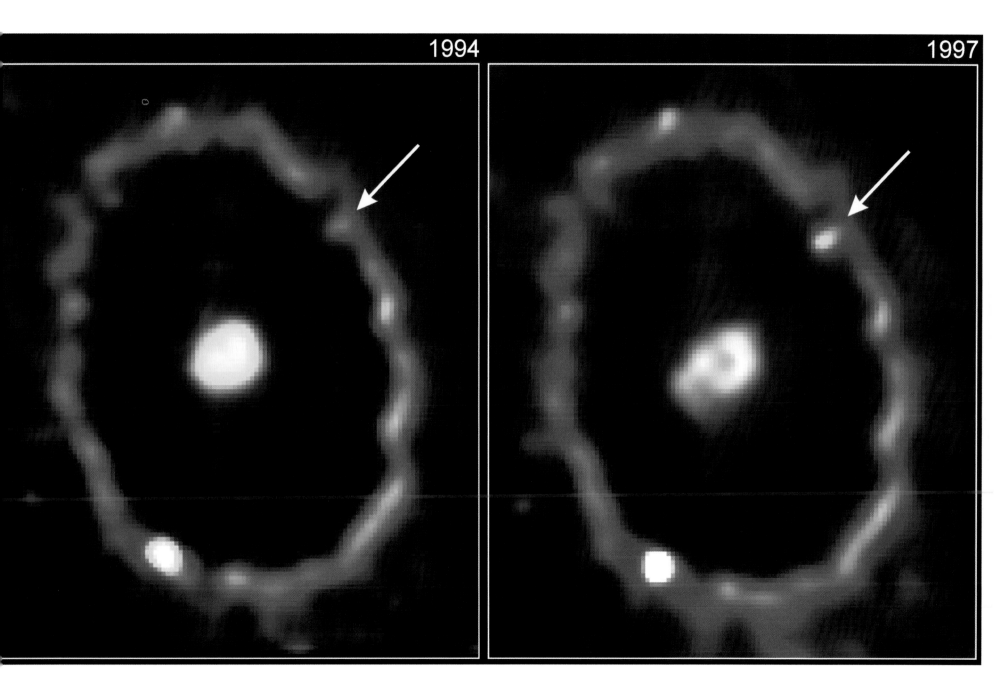

1994

1997

A bright spot has appeared on the innermost ring around the supernova. The first extensions of fast-moving gas from the explosion have reached the ring and heated an area of it. Such hot spots are expected to appear all around the ring. (Garnavich and NASA) of the ring had been visible before, so it appears reasonable that the fastest expanding gases hit this spot first. In only a few years, the entire ring will be covered with such hot regions. Model calculations show that the inner ring should increase in brightness by a factor of 1000 as the main parts of the supernova ejecta collide with it. Although it will not become visible to the naked eye, as the supernova outburst itself did, the heated ring will be within easy reach of smaller telescopes on the ground and in space. At the time of this writing, it is an object of only 19th magnitude, but during its "second wind," it could easily reach 13th magnitude again. The main impact of the ejected materials from the supernova onto the inner ring is not expected before 2005.

The chemistry of the inner ring around the supernova is revealed by Hubble's STIS instrument. The entire ring could be observed at once in the spectrograph's entrance aperture and was dispersed into its various colors. Because the gas of which the ring is composed glows in discrete lines, numerous images of the ring appear next to one another. Oxygen (green), nitrogen and hydrogen (orange), and sulfur (red) can be seen. (Sonneborn & Pun and NASA)

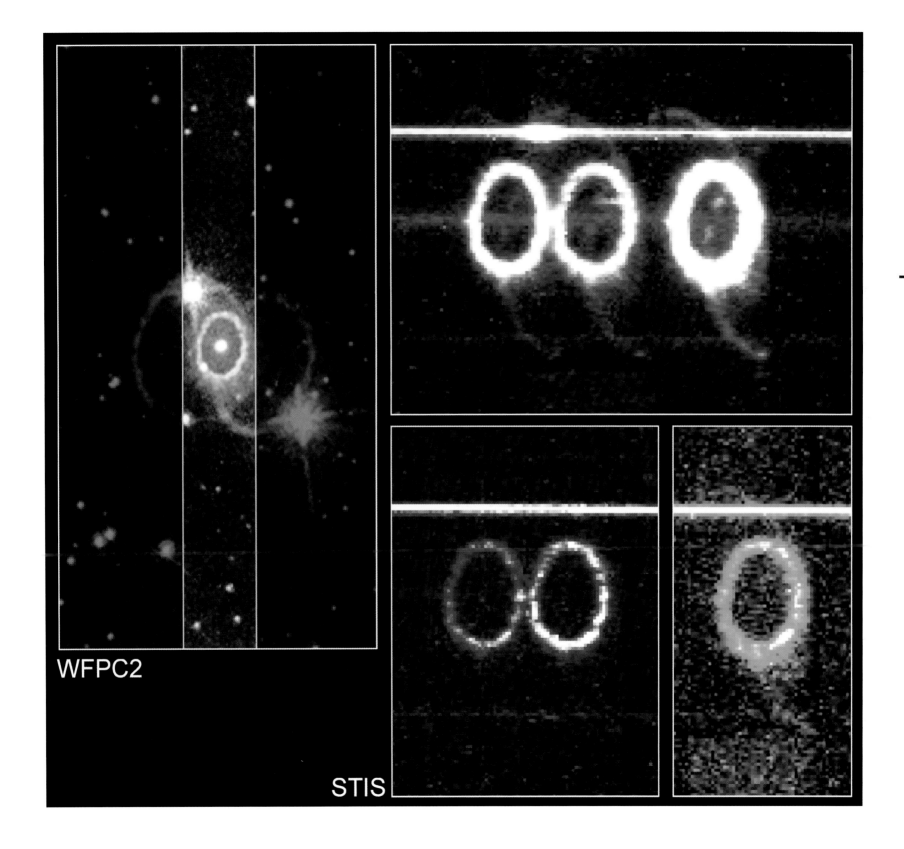

WFPC2

STIS

Neutron Stars

The stellar graveyard is populated by white dwarfs, neutron stars, and even black holes. In close binary systems, these objects can be detected easily by their interactions with other stars—either by their orbits or as dwarf novae, classical novae, and x-ray bursters. But single stars are much harder to identify because they reveal themselves only by their own faint luminosity. We have identified a fairly large number of white dwarfs and a few rapidly rotating neutron stars. These remnants of supernovae that occurred a few centuries or millennia ago are the pulsars, which announce themselves by regular "ticks" in the radio domain. The most famous pulsar may be an inhabitant of the Crab Nebula: several interesting phenomena resulted from a supernova in 1054. The Crab Nebula—with its bizarre structure consisting of the shell of the exploding star with the pulsar in its center—remains the target of many investigations. Observers on the ground knew for quite some time that the Crab Nebula can exhibit change on time scales of months or years, and its overall expansion had also been detected quite early.

But nobody expected the dynamic events witnessed by the Hubble Space Telescope. Near the pulsar, we can observe patterns of stripes that continuously change shape! The young neutron star constituting the pulsar rotates on its axis about 30 times per second. Its strong magnetic field in combination with the rapid rotation serves as a particle accelerator. The changing luminous patterns may be a result of the interaction of this "pulsar wind" of

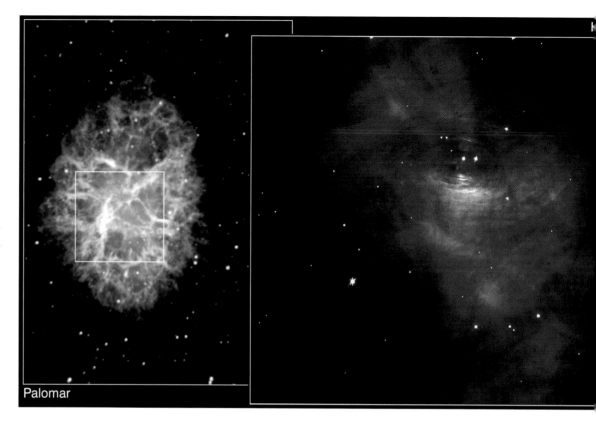

Palomar

particles near the speed of light with the gas of the nebula. One model postulates the existence of two focused jets along the axis of the pulsar, in addition to the acceleration of particles in the equatorial plane. A particularly intense spot that frequently changes location may be caused by a shock front from one of the polar jets. The other polar jet, pointing in the opposite direction, remains invisible. Until recently, the pulsar had been expected to emit its wind in all directions; now it appears more likely that particles are accelerated to escape the magnetic field preferentially in the equatorial plane. Along the rotational axis, the twisting of the magnetic field lines provides the required acceleration.

What does an old, isolated neutron star look like? Several hundred million of them should exist in our galaxy. Such an old neutron star should appear as a relatively hot but faint object. In 1992,

Hubble reveals the dynamics of the Crab Nebula, including the interaction of the pulsar with the remnants of the stellar explosion of 1054. The dense corpse of a star rotates about 30 times per second and acts as a particle accelerator, causing rapidly changing illumination patterns in the gas of the nebula. (Hester & Scowen and NASA)

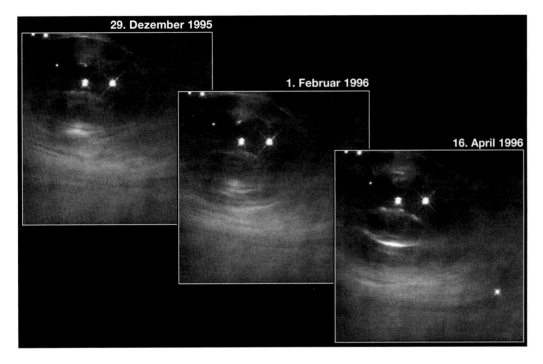

29. Dezember 1995

1. Februar 1996

16. April 1996

ROSAT observed a bright x-ray source for which no optical counterpart could be found at first. Hubble finally correlated this source with a faint blue star of 25th magnitude. Because it is located in front of an interstellar molecular cloud, its distance can be estimated: it is at most 400 light years away. The brightnesses in the optical, ultraviolet, and x-ray ranges yield the approximate temperature and size of the star: a sphere of at most 28 kilometers diameter, with a surface temperature of about 650,000 K. Knowing the size of the star is important for theoretical investigations into the state of the dense matter: it is so densely packed that electrons and neutrons are compressed into a strange form of neutron liquid. This is the densest known form of matter—a handful has as much mass as an aircraft carrier!

Hubble also successfully hunted for other isolated neutron stars. The Faint Object Camera provided an ultraviolet image of the pulsar 1055-52, one of the most efficient sources of gamma radiation known. It emits half its energy in this domain, and only very little energy goes into the visible range. In addition, a star 100,000 times brighter than pulsar 1055-52 is located right next to it, making observations with ground-based telescopes impossible.

With this observation, the Hubble Space Telescope had succeeded in investigating four of the eight pulsars that are known to emit visible light—including the Crab pulsar and three other old neutron stars, where the supernova remnants are no longer in evidence. The observations of isolated neutron stars are particularly important for investigating the mechanisms involved in pulsar phenomena. Only seven cases without the characteristic radio emission are known, and only one of them emits gamma rays, while all of them are strong x-ray sources.

A Festival of Colors and Shapes

Once the nuclear fusion reactions of hydrogen to helium have exhausted the hydrogen and left nothing but helium in the core of a normal dwarf star like our Sun, the star begins to develop into a red giant. When enough helium is available in the core, the violent fusion of helium to carbon and oxygen begins. At this point, the star becomes brighter and redder and begins to lose matter into surrounding space. More and more violent and short-lived fusion cycles produce heavier and heavier elements, up to iron. These cycles cause the star to shed more and more of its outer layers into space. Finally, all nuclear fuel is exhausted, and the star becomes a white dwarf, drawing energy only from its slow shrinking process. White dwarfs exhibit high temperatures, both in their interiors and on the surface. As soon as the surface temperature reaches more than about 10,000 K, the light is energetic enough to induce the previously emitted matter to glow.

This fluorescent gaseous material forms what is called a planetary nebula. These objects, which come in a multitude of shapes and colors, are signs of the late stages in the evolution of stars. They have nothing at all to do with planets; the misleading name arose because some of these objects resembled the images of distant planets such as Uranus or Neptune when viewed through the first telescopes. Planetary nebulae are the gravestones of stars that can no longer generate energy from nuclear processes. Once the central star, which may have reached temperatures of several hundred thousand kelvins as a white dwarf, exhausts even its gravitational energy and cools below 10,000 K again, the planetary nebula should cease to glow. After such a long time, the gas has totally dispersed into interstellar space and mixed with interstellar gas, where it may later be recycled in a new generation of stars.

The Stingray Nebula, Henize 1357, may well be the youngest planetary nebula: only 20 years ago, its gas was not hot enough to emit light! But because the temperature of the central star has increased rapidly, astronomers can watch the formation of a planetary nebula "live": the entire transition phase may take only about 100 years. Because Henize 1357 is unusually small and quite far away—18,000 light years—only the Hubble Space Telescope could resolve the bewildering structures that already characterize the shape of this nebula. One of the new findings involves the two large gas bubbles, which are a frequent phenomenon in planetary nebulae. A dense ring around the dying star lets gas from the star escape only along its axis, but not through the ring itself; the ring acts as a nozzle. It was also discovered that the central star was not a single star but a binary system. The companion star may have a significant influence on the shape of the entire nebula, and it could be responsible for the gas ring. The nebula's two axes of symmetry may also be a result of the binary nature of the central system.

The great zoo of planetary nebulae. Most of the images were taken by WFPC2; some were taken by NICMOS. The Hubble Space Telescope shows the splendor of these nebular structures, which dying relatives of our Sun leave in space and then induce to glow.

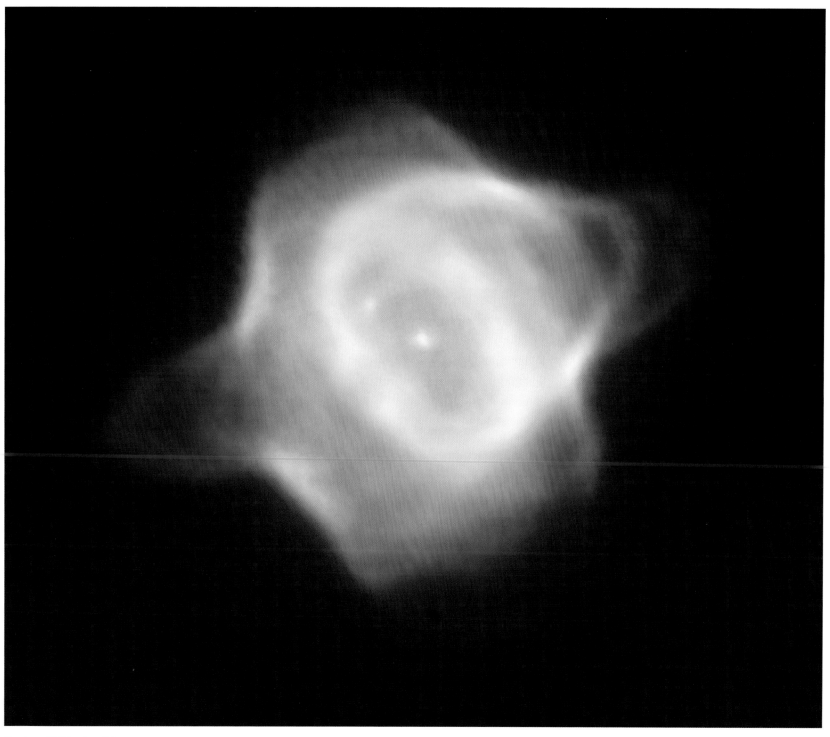

**Henize 1357, the Stingray
Nebula. (Bobrowski and
NASA)**

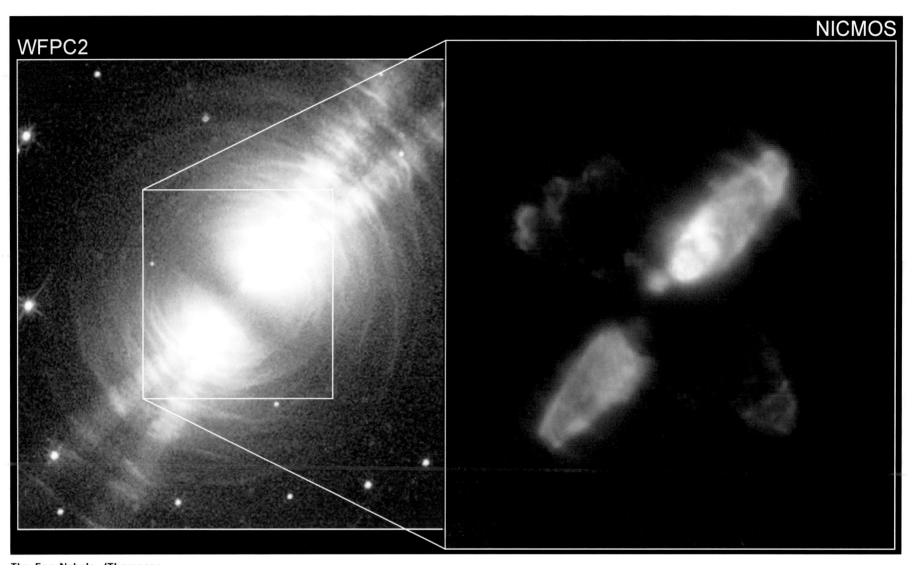

WFPC2

NICMOS

The Egg Nebula. (Thompson
et al. and NASA)

Left: CRL 2688, the Egg
Nebula. (Sahai & Trauger and
NASA)

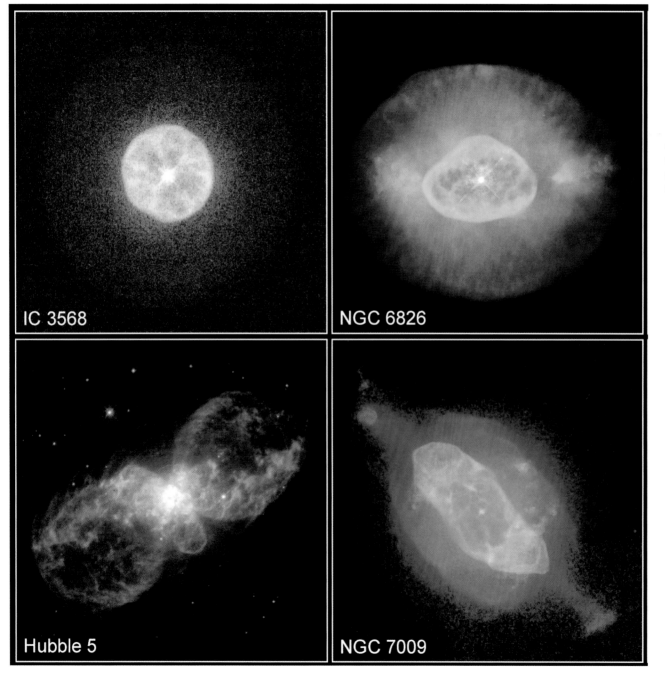

A gallery of planetary nebulae. (Left: Bond et al. and NASA. Right: Kwok et al. and NASA)

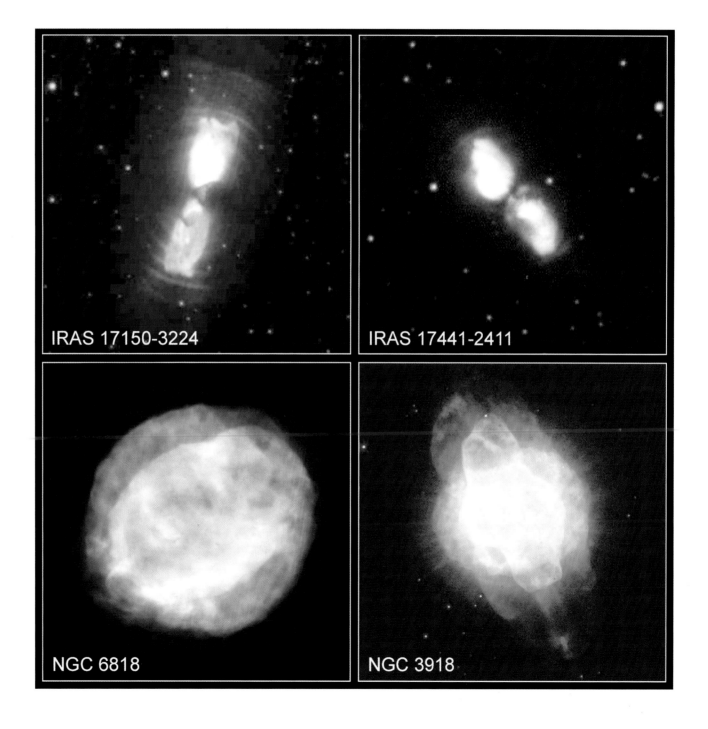

IRAS 17150-3224

IRAS 17441-2411

NGC 6818

NGC 3918

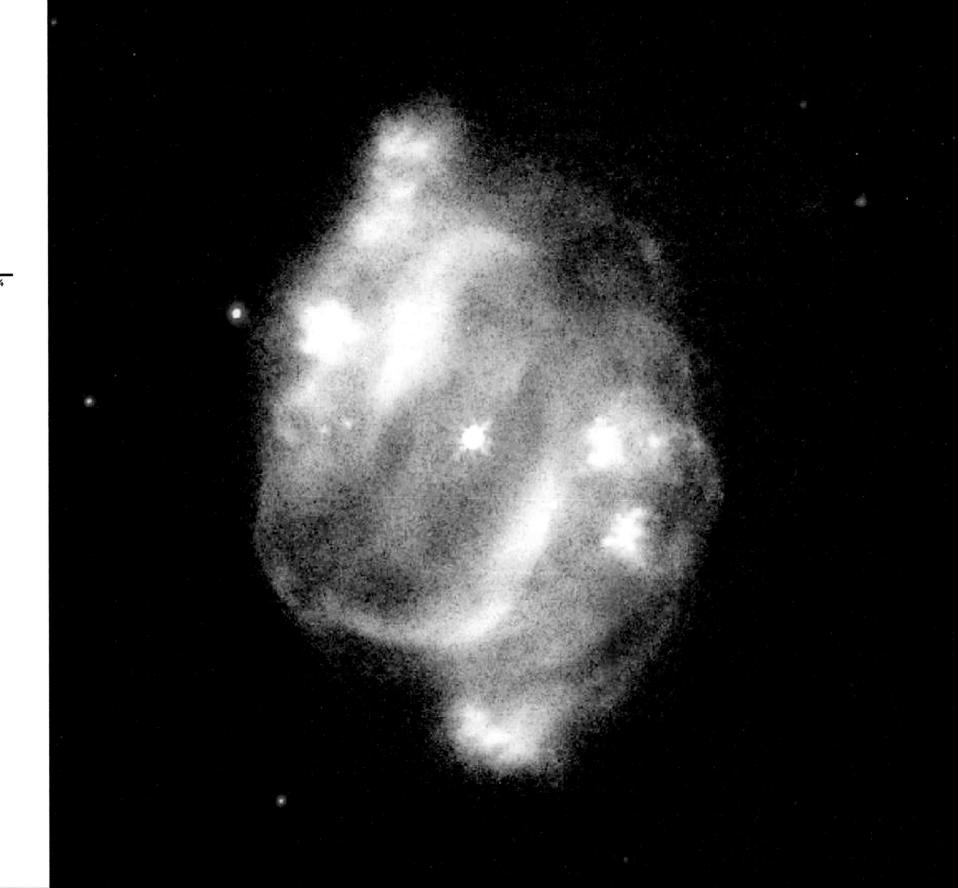

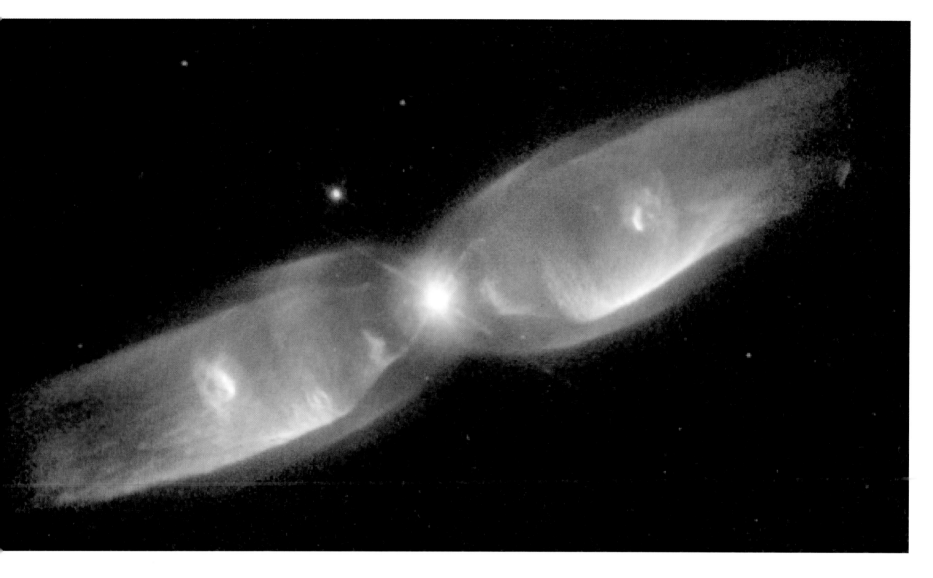

M 2-9, the Butterfly Nebula.
(Balick and NASA)

Left: NGC 5307. (H. Bond et
al. and NASA)

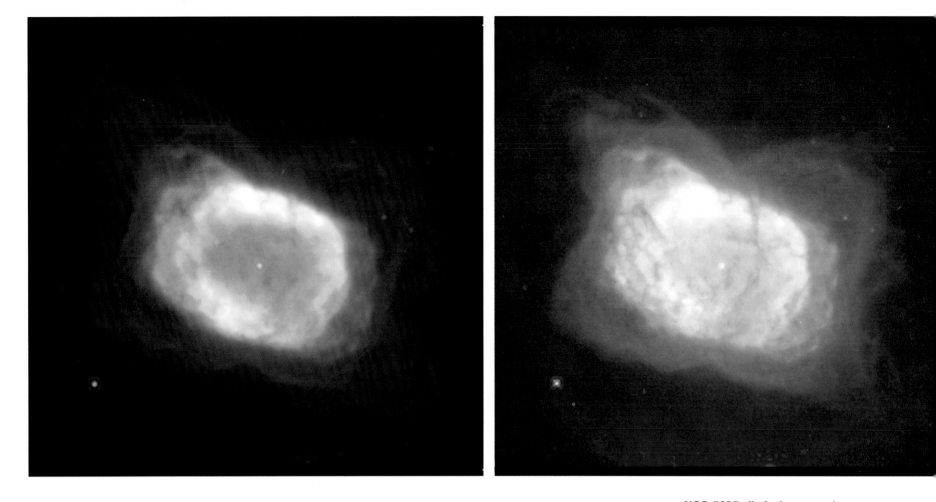

NGC 7027. (Left: Latter and NASA. Right: Bond and NASA)

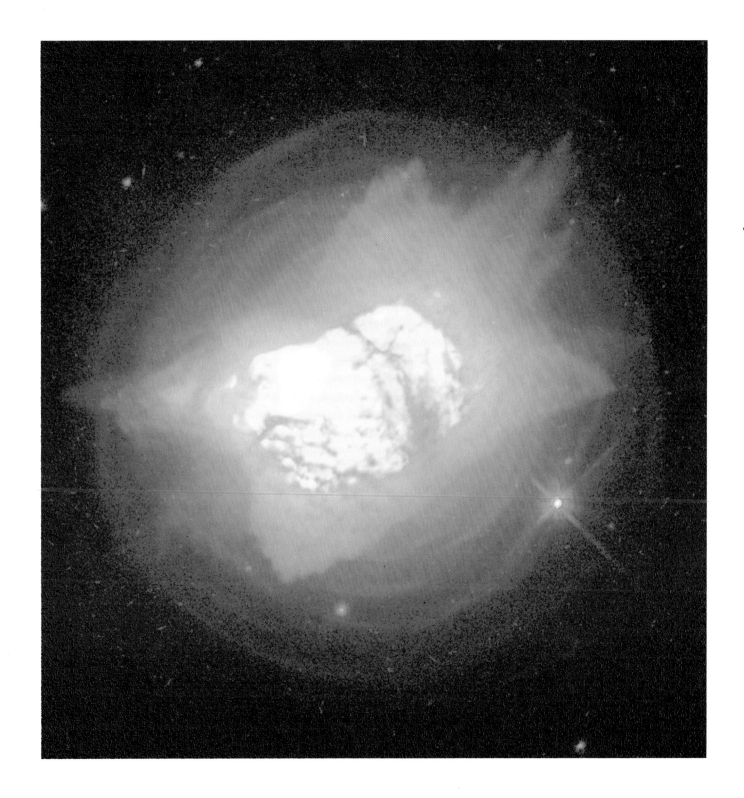

NGC 7293, the Helix Nebula.
(O'Dell and NASA)

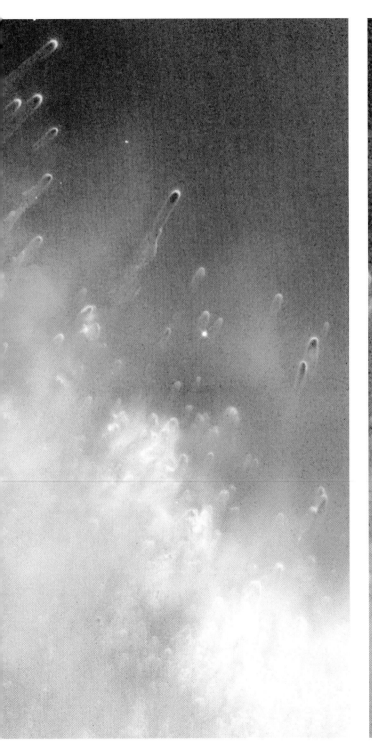

MyCn 18, the Hourglass Nebula. (Sahai & Trauger and NASA)

The Egg Nebula, CRL 2688, represents a slightly later point in the development of a planetary nebula—in this case, the appearance of the nebula may be a few hundred years in the past. The arcs in the structure of the nebula are interpreted as shells ejected by the dying star in episodes every 100 to 500 years. Hubble's new NICMOS camera added more information: the infrared shows a pronounced dumbbell structure of molecular hydrogen. Here, matter from the dying star in the center streams out along the polar axis at a speed of 100 kilometers per second; it collides with previously ejected material having a velocity of "only" 20 kilometers per second. This also explains the strange shape of the Egg Nebula in visible light: the matter in the two streams darkens the centers of the cones of light emanating from the interior of the nebula.

What do older planetary nebulae look like, and what can they tell us about the final stages of stellar evolution? Over the years, Hubble has assembled an amazing collection of colors and shapes of planetary nebulae. With its superior resolution, the WFPC2 shows known structures more clearly than ever and brings new ones to light. These images illuminate the complexity of the processes that take place during the last phases in the evolution of a star like our Sun—processes that may be influenced by invisible companions such as small stars, brown dwarfs, and large planets. We see torus-shaped dust regions focusing

the ejected gas, isolated bubbles and narrow rays of denser gas, and symmetric structures on both sides of the central star. The basic model for the origin of a planetary nebula postulates collisions of winds of different speeds that the dying star has discharged more and more rapidly. These winds produce a shell that is ultimately induced to glow by the ultraviolet light from the white dwarf in the center. The multitude of details makes it clear, however, that many aspects of this process are not yet fully understood.

Another fascinating example is MyCn18, a planetary nebula at a distance of 8000 light years that was discovered by astronomers Margaret Mayall and Annie Cannon. Hubble's image shows the shape of an hourglass whose walls display peculiar patterns. Models for the formation of planetary nebulae predict such hourglass shapes as a result of the expansion of a faster wind into a previous slower one, the slower one being denser at the equator than at the poles. Surprisingly, the central star lies in a structure with an axis of symmetry that deviates radically from that of the outer nebula. Such a difference is hard to understand. Next to the center, two small elliptical rings appear, with the central object at their intersection. The patterns in the outer shell may have been caused by individual episodes of ejection of matter or by a dense stream of matter interacting with the walls of the hourglass. Gravitational effects from a companion star may also cause some of the irregularities.

The planetary nebula NGC 7027, at a distance of 3000 light years in the constellation Cygnus (the swan), appears to be an extremely intricate structure. The central star can be seen as a faint dot in the center, and the bright inner region is surrounded by a complex network of red dust clouds. A number of concentric, faint blue shells encircle the nebula. A weak, episodic stellar wind may have generated the outer shells. Later, the dying star lost all of its outer layers simultaneously in a catastrophic loss of mass called a "superwind"; this created the bright, inner region of the planetary nebula. The late phases of mass loss were not symmetric, and localized dense gas clouds formed. Hubble's NICMOS camera unveiled the presence of a cool molecular gas around the hot luminous one for the first time. NGC 7027 turned out to be a young (less than 1000 years old) planetary nebula in the midst of development.

The Butterfly Nebula is a typical bipolar planetary nebula with a binary system in its center. One of the stars has probably drawn mass from the other, forming an accretion disk with 10 times the diameter of Pluto's orbit around the Sun. This disk forces the gas, ejected from one of the stars at more than 300 kilometers per second, into two symmetrical bubbles, similar to the action of a nozzle in a jet engine. As ground-based observations have already shown, this nebula—at

a distance of 2100 light years—expands over time; it existed for only about 1200 years. Hubble 5, at a distance of 2200 light years, is another bipolar planetary nebula.

NGC 6826, the "blinking planetary," contains classic examples of "fast, low ionization emission regions" or FLIERs—two groups of deeply red points of gas ejected by the dying star about 1000 years ago. The regions consist mainly of nitrogen. It is still unclear whether the FLIERs emanate from the polar axis or from the equatorial plane of the central star, but the high degree of symmetry is surprising in any case. NGC 7009, also called the Saturn Nebula because of its characteristic, handle-shaped FLIERs, shows particularly well how the FLIERs are connected to the central star by narrow jets or rays of matter. NGC 5307 has a surprising degree of central symmetry, with every gas bubble having its equivalent on the other side. Hubble also provided images of comparatively simple and mainly spherical planetary nebulae. IC 3568 displays a bright inner shell with sharp edges and a fainter, diffuse outer envelope. And NGC 3918 consists of a roughly spherical shell containing an elongated inner balloon; here, the fast stellar wind attempts to break out of the shell.

A few planetary nebulae show yet another exotic phenomenon: "cometary" knots. In the Helix Nebula, NGC 7293, located in the constellation Aquarius (the water carrier) at a comparatively close distance of 450 light years, strange comet-shaped structures are visible even on images taken from the ground. The Hubble images show about 3500 of these objects, but their total number may be even higher. Each of these gaseous structures has a "head" of more than twice the size of our solar system and a "tail" longer than a thousand times the distance from Earth to the Sun. The knots have photoionized surfaces, resulting from bombardment with energetic radiation. But their insides contain large amounts of neutral gas. A typical "cometary" knot has a mass 1/30,000 that of the Sun; taking into account all 3500 visible structures results in a total of about a tenth of a solar mass—scarcely less than the ionized gas in the entire nebula! These strange knots are therefore an important component of this planetary nebula and are probably a general phenomenon in these stellar gravestones. But where do they come from, and where are they going?

The knots can probably be explained by the interaction of a relatively recent hot gaseous wind from the central star with the denser, cooler material in the planetary nebula, which was ejected some 10,000 years ago. This interaction may have led to instabilities that caused these astonishing comet-shaped objects. Computer simulations have not yet provided sufficiently accurate models, however. Perhaps the condensation of these knots is aided by the presence of "seeds," objects that already existed during the lifetime of the star. Did they remain from the star's formation, or are they

a cloud of large, real comets? Many questions about these strange objects remain to be answered in detail. And what happens to these knots in the future? They contain enough neutral gas to withstand the stress of the planetary nebula phase and to remain as compact, maybe even solid, objects. Such a process may have occurred in a large number of old planetary nebulae. In that case, interstellar matter would consist of yet another important—but largely invisible—component.

If every dying star produces these objects, they could even account for a substantial fraction of the "dark matter" of the universe, for which cosmology is searching. Planetary nebulae have another important role: together with supernova remnants, they generate heavier elements in the galaxy. Depending on their size, the dying stars eject various amounts of carbon, oxygen, iron, and other elements into space; life as we know it would be impossible without these heavier elements generated in earlier generations of stars. The variety of planetary nebulae reflects the multitude of stars, to which we owe our very existence. Planetary nebulae also show us what our solar system will look like in a few billion years. Undoubtedly, we could not survive the death throes of our Sun, but let us imagine that we could return after things had stabilized again. We would marvel at the interior of a planetary nebula, perhaps as complex as those seen in Hubble images. The night sky would show colors, jets, bubbles, rings, and spots of light. It would be a spectacular view.

Old Couples

Dwarf novae, classical novae, recurring novae, symbiotic stars: the common denominator of these objects is their binary nature—two stars circle each other with periods of hours to many decades. In these binary systems, one of the stars is a white dwarf: a star the size of Earth but with the mass of the Sun that evolved from a red giant and has now exhausted its nuclear fuel and become a slowly cooling sphere of helium, carbon, and oxygen. After millions of years, it will reach temperatures too low for light to be emitted, and it will vanish from our view as a black dwarf. If such a star has a close dwarf companion, its gravitational pull can draw hydrogen-rich matter from the outer layers of the companion. In the case of a giant companion with its usual stellar wind, the white dwarf could accrete matter from the wind. This matter will not simply fall into the star, however, because its angular momentum would prevent it from doing so. Instead, the matter will accumulate in an accretion disk around the white dwarf, from which it will slowly rain down onto the star's surface. Like oceans of water on Earth, an ocean of hydrogen will form on the surface of the white dwarf. And hydrogen can be used as nuclear fuel.

Once a critical amount of hydrogen has accumulated on the surface of the dwarf star, and the density on the bottom of the "ocean" has reached about 10,000 times that of water, nuclear fusion sets in explosively, ejecting the outer layer into space at velocities of several thousand kilometers per second. During this explosion, the brightness increases by factors of 1000 to 100,000. Early astronomers erroneously called the newly appeared bright star a "nova," short for the Latin "stella nova," or new star. The name persisted, even though the phenomenon is the sign of old age in a star. The Hubble Space Telescope observed such a classical nova in detail. The nova Cygni 1992, or V1964 Cygni (the official name in the catalog of variable stars), was discovered by an amateur astronomer in the early morning hours of February 19, 1992. The nova was observed intensely with telescopes on the ground and in space. A direct image was taken by Hubble a little less than two years after the explosion: the ejected shell, showing a ring-shaped structure with a diameter of 163 billion kilometers, is clearly visible. Classical novae are expected to repeat their outbursts every few centuries. The wait is much shorter for so-called recurring novae, with typical intervals between eruptions of 10 to 100 years.

T Pyxidis is a peculiar recurring nova in the southern constellation of the Ship's Compass, at a distance of about 6000 light years. Since its discovery, it has erupted five times—in 1890, 1902, 1920, 1944, and 1966—and astronomers are longingly waiting for the next event. The frequent recurrence of these outbursts and the large amounts of ejected matter make T Pyxidis an interesting case. Periodically ejected shells interact with one another, leading to concentrations of matter. More than 2000 such clumps, each the size of our solar system, cover an area about a light year in diameter. Only one

Erde

HST

Legacy of a nova. The recurring nova T Pyxidis (shown in a ground-based image on the left, and an HST image on the right) has surrounded itself with a cloud of about 2000 small gas bubbles that were ejected into space during several episodes of explosions. The collision of gas shells with different expansion velocities may have led to instabilities and the formation of these bubbles. (Shara et al. and NASA)

other nova is known to possess such a complex shell: GK Persei, a nova that appeared in 1901. This nova may have ejected its shell into the material of the surrounding, fossil planetary nebula. It had originated when a red giant in the GK Persei binary system developed into the white dwarf that we now observe. Because this planetary nebula shell, which has remained visible for several hundred thousand years, can still be observed in GK Persei, this must be an extremely young nova. Most of the other 200 known novae in our galaxy are much older.

The combination of white dwarf and red giant occurs frequently in recurring and symbiotic novae. If the red giant star shows a particular form of variable brightness, it is classified as a Mira variable. The prototype of this group of stars, Mira in the constellation Cetus (the whale), also known as Omicron Ceti, was discovered by David Fabricius on August 13, 1596. It pulsates with a period of 332 days and ejects large amounts of gas and dust in its strong stellar wind. Several decades ago, the existence of a white dwarf companion for Mira was established. This white dwarf circles the giant at 70 times the distance from Earth to the Sun—close enough to collect matter from the giant star. Mira's distance from us amounts to about 400 light years. With its Faint Object Camera, Hubble has obtained images of this system in optical and ultraviolet light. They clearly show the two stars and even give indications of their interactions. Once the highest possible resolution is achieved with the help of computer-aided image reconstruction, the ultraviolet image shows a small hook-shaped structure in the direction of the companion. Is this matter from Mira, pulled by gravity to the white dwarf, or material in the atmosphere of Mira, heated by the white dwarf?

In visible light, Mira appears as an ellipsoid; that is, slightly flattened from the usual spherical shape of a star. Its measured angular diameter

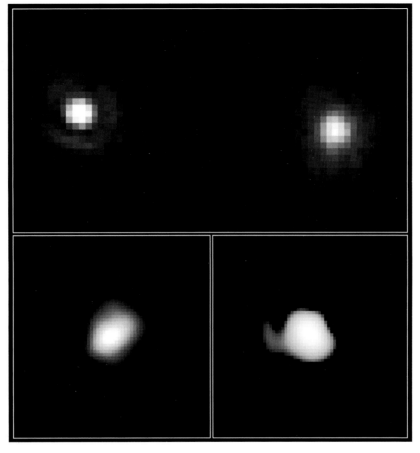

The cool giant Mira (Omicron Ceti) up close. In the top image, the FOC separates the star (right) from its close companion at a distance of only 0.6 arc second, or 10 billion kilometers. On the bottom, Mira is resolved into a small disk by image processing techniques. In visible light (bottom left), the star appears oval, rather than spherical. In the ultraviolet (bottom right), a small gas cloud points into the direction of the companion star. (Karovska and NASA)

amounts to 0.06 arc second, corresponding to 700 solar diameters. If Mira were in the Sun's place, it would include the orbits of Mercury, Venus, Earth, and Mars; its surface would extend to two-thirds of Jupiter's distance. The Hubble Space Telescope also observed the shapes of several relatives of Mira. These observations were made with the Fine Guidance Sensors, which are routinely used for the precise orientation of the satellite in space. These instruments determine the positions of stars in their field of view extremely accurately with the help of interferometric methods; they achieve an angular resolution of about 1/100 of an arc second. These instruments provided information about the nonspherical shape of two more Mira variables. The image of R Leonis has an angular dimension of 0.07 by 0.08 arc second and that of W Hydrae an angular dimension of 0.08 by 0.09 arc second. For both stars, this translates into a size of approximately 1.3 by 1.45 billion kilometers.

With these measurements, Hubble has provided valuable insights into the shapes of individual stars. But as we have seen, Hubble's exploration of the realm of stars in our Milky Way ranges from their birth in stellar nurseries to their evolution as single or binary objects, and to their lives' spectacular end products of planetary nebulae and neutron stars.

Part 4

Planets

Planets around Other Stars

The Case of Beta Pictoris

The Hubble Space Telescope was not provided with any special equipment to search for planets around other stars, but a number of its observations have contributed to this very popular topic. The discovery that every other young star in the Orion Nebula is surrounded by a disk in which potential planets could develop provided cause for optimism. The next step in the search for extrasolar planets is much more difficult: only two dust disks around older stars have been discovered in visible light. The disk around the star Beta Pictoris is the more impressive case. Since its discovery in 1984, it has been examined by several cameras and telescopes on the ground as well as by the Hubble Space Telescope. The first results from HST showed the disk to be even thinner than anticipated from ground-based images. This thinness indicates that the disk—which we see almost exactly edge-on—must be fairly old; the dust has had time to collect. In such a thin disk, it is very probable that larger clumps have already begun to emerge.

Our own solar system may have looked much like this when comets and planets were just beginning to form. It is possible that at least the first phases of planetary formation are under way in Beta Pictoris. The mere existence of the disk is an argument in favor of this hypothesis—if a process of planetary formation were not under way, the disk should have dissipated from frequent collisions of its dust particles. Might there even be grown planets hiding in this 300-billion-kilometer-diameter dust disk? A WFPC2 observation showed that the inner parts of the disk are warped. A planet of roughly Jupiter's mass, in an orbit with a slight inclination to the plane of the disk, could induce such warping. It is next to impossible, however, to deduce the characteristics and orbit of a potential planet from the observed warp of the disk. The radius of the planet's orbit could range from 150 million to 4.5 billion kilometers, and its mass could be from 1/20 to 20 Jupiter masses. An alternative explanation is that a chance encounter with a nearby star might have warped the disk temporarily.

When the disk was finally observed with near-optimal equipment, STIS in imaging mode and with a special aperture that covered the central star itself, the new images did not shed much more light. The warp could be measured even closer to the star, but now the data allowed three different interpretations. The warp in the disk could be caused by a hypothetical planet, a brown dwarf in a much larger orbit around Beta Pictoris, or the chance passage of another star through the area. At least the possibility that radiation pressure from Beta Pictoris itself might have caused the warp could now be excluded.

Brown Dwarfs and Giant Planets

A few very faint companions of other stars can be observed directly: Hubble images show a faint point

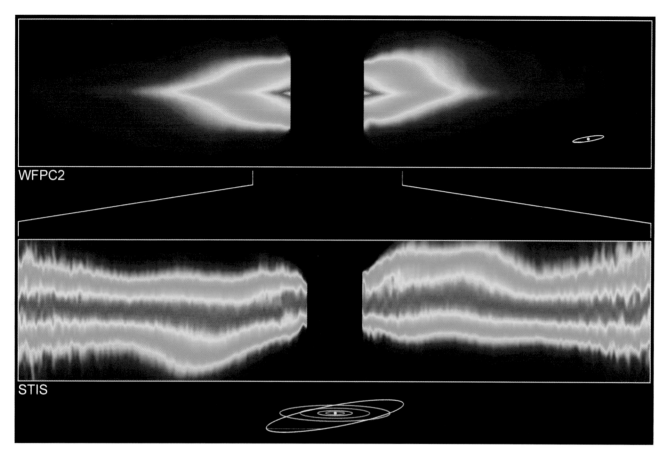

WFPC2

STIS

The warped gas disk of Beta Pictoris. Have WFPC2 (left and top right) and STIS (bottom right) really discovered the effects of a planet or other companion? The WFPC2 observations of the faint disk next to a bright star were difficult, but STIS contains a special diaphragm to cover the bright star. The disk and its considerable warp can be observed closer to the star. (Burrows & Krist, Schultz, Heap, and NASA)

of light next to our nearest neighbor, Alpha Centauri C (also called Proxima Centauri). It is suspected that this faint point moved slightly over a period of three months. If these observations at the edge of detectability are more than a camera artifact and stand the test of further analysis, which is uncertain at the time of this writing, they would indicate either a planet of at least 10 Jupiter masses or a brown dwarf. Other methods have failed to find a companion of Proxima Centauri; the spectral lines do not show the periodic shifts that would be induced by the revolution of a planet around the star. A rhythmic but minute motion of the star under the influence of a large planet is exactly the method that has been

used to discover the majority of all known planets around other stars. Maybe the orbit of the companion is very elliptical, so that the effect is observable only for short periods of time; but theoreticians regard this possibility as highly unlikely. The investigations of this star continue and should soon lead to clearer results.

The First Image of an Extrasolar Planet?

At the end of May 1998 another picture of a putative planet was presented to the public. The image had been captured in mid-1997, but its importance had been recognized only recently. So far, all the possible

The binary star TMR-1A/B (top right) and its apparently escaping protoplanet (top left) connected by a 100-billion-mile long gaseous tail. (Source: S. Treby and NASA)

planets circling distant stars — barely ten — have been discovered by indirect methods; they are too distant from us and too close to their stars to have any chance to be seen directly by Hubble. This new candidate may be an exception: It is in a much more visible, albeit unexpected, place, some 1400 times as distant from its parent as the Earth is from the Sun. This object, TMR-1C, is also, strictly speaking, not a planet, since it is no longer circling the presumed central star, the binary star TMR-1 A and B, but seems to have escaped its gravity. All three objects are in the constellation Taurus, in a star-forming region called the Taurus molecular ring; their age is estimated to be only a few hundred thousand years.

Current speculation assumes that TMR-1C was formed at the same time as the two stars A and B; its orbit around the two was apparently so convoluted that it was finally catapulted into interstellar space (much as the Vanguard and Pioneer satellites have been sent out of the Solar system by their encounters

with Jupiter and other planets). If the three objects are indeed the same age and are related, one can also conclude that the mass of TMR-1C is on the order of two to three times the mass of Jupiter, that it radiates its own light (in the infrared only) because it is still contracting and being heated by the contraction. On the other hand, if TMR-1C is a much older object, it could be a much more massive brown dwarf rather than a Jupiter-size planetary object. There is, further, a roughly 2% chance that TMR-1C is only accidentally in the field of view and is not a part of the TMR-1 system at all; in that case it might simply be a more distant ordinary star.

There are plans to use the Keck telescopes to make spectroscopic observations of TMR-1C to clarify its nature; the glowing filament is also being examined. Additional studies are underway to determine the proper motion of the object, to see if it is indeed moving away from the binary system in the expected direction.

If it turns out that we are indeed dealing with a protoplanet, it will have fundamental implications for astrophysics, and possibly for theories about the formation of our own Solar System. According to the standard models of planet formation, such an object could not exist, since gas clouds take millions of years to condense into protoplanets, so that a Jupiter-size object only a few hundred thousand years old would not be possible. There are of course alternative hypotheses. According to one of these, a gas cloud could collapse into a protoplanet almost

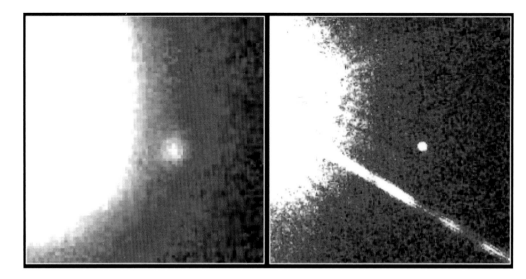

instantaneously, in a few hundred years after the onset of an instability in the gaseous disk surrounding a very young star. Only such a mechanism could have produced a planet as young as TMR-1C seems to be.

The classification boundaries between giant planets and brown dwarfs, and between these "failed stars" and the smallest true stars, are difficult to establish. Several faint points of light in the sky have moved from one category to the other, based on observations of new characteristics of the objects. The Hubble Space Telescope can often provide important new details. A case in point is the brown dwarf Gliese 229B. Although it was detected from the ground, only Hubble could investigate it as an individual object. Gliese 229B is the companion of an ordinary star and has a luminosity of only two- to four-millionths that of the Sun. But this faint speck of light was tracked down in its orbit around yet another star! Its spectrum, resembling that of Jupiter, discloses that it is a brown dwarf. A large amount of methane exists on this object, which in turn tells us that the surface temperature is 600°C to 700°C. In addition, GL 229B's atmosphere contains more water than has ever been seen in any true star as cool as this one. GL 229B's mass is 4 to 6 percent of the Sun's mass.

In contrast, the object Gliese 106C is definitely a small star, forming a binary system with a much brighter star. Hubble was able to provide images of the very faint companion next to the blinding main star, and it allowed more accurate measurement of the binary system's orbit. The mass of Gliese 106C is 8 or 9 percent of the Sun's mass. This is is only a small amount above the lower limit for energy production from nuclear fusion. The luminosity of this starlet is correspondingly small. If it were in place of the Sun, it would only be four times brighter than the full Moon. And its surface temperature is just 2300°C.

The brown dwarf Gliese 229B. This object was detected with a 1.5-meter telescope on the ground (left), but Hubble provided a much sharper image (right). The small companion of Gliese 229B has a mass only 20 to 50 times that of Jupiter. Perhaps a space mission in the next century, such as one of those currently being planned by NASA, will take a similar image—but of a true planet like Earth. (Nakajima et al. and NASA)

At the Edge of the Solar System: Pluto and Trans-Neptunian Objects

The first images showing details of Pluto's surface, a result of extensive analysis and processing. The original Hubble images (small images on top) cover only a very few pixels. The large detailed image resulted from their superposition by computer. A map of Pluto was first produced from the single images and then projected onto a sphere. (Stern & Buie and NASA)

Pluto

The most distant planet of our solar system, Pluto, was not discovered until 1930. Even for Hubble, it has been a tough nut to crack. With an angular diameter of only a tenth of an arc second, it extends over only a few tens of pixels in Hubble's cameras. Before the installation of corrective optics, it remained a blob of light without any detail, accompanied by a second, fainter blob, its moon Charon. But soon after the installation of COSTAR, several images of Pluto were taken with the Faint Object Camera that showed bright and dark areas for the first time. The result was published in the astronomical literature, but mainly as proof of the quality of the camera with its new corrective optics. Two years later, the results of a complex analysis of a different, larger set of Pluto images, also taken with the Faint Object Camera, became available. The various bright and dark pixels from each individual image were combined into a comparatively detailed map of Pluto.

In these images, Pluto shows a bright polar cap in the north. The body of the planet is covered by a network of dark and bright areas in a pattern that has not been seen on the various other icy planets of our solar system. Unfortunately, the images do not reveal the physical nature of these structures. They could be lava-filled basins or fresh impact craters (as on the Moon), or we may be seeing an irregular layer of frozen material from Pluto's atmosphere. The brightest areas reflect light, much as snow does.

Doubts about Trans-Neptunian Objects

Unfortunately, we may have to bid farewell to one of the more exciting discoveries from Hubble observations at the edge of the solar system. It is very likely that the attempt to take images of numerous comet cores in the Kuiper belt failed. It would have

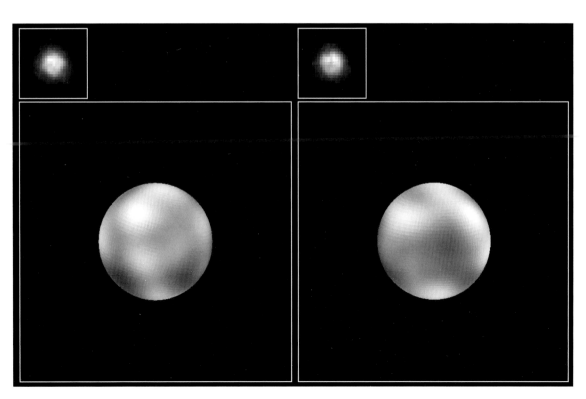

been nice to close the gap: outside of Neptune's orbit, there is only one large object, Pluto itself, with a diameter of 2300 kilometers. But there should be some 10,000 trans-Neptunian objects with diameters of a few hundred kilometers, of which only about 60 have been observed. And these same far reaches of our solar system should contain millions of comet cores. It was assumed that at least those objects the size of comet Halley (10 kilometers in diameter) or larger would be detectable by the Hubble Space Telescope. In 1995, Hubble obtained images that seemed to show the comets. Individual comet cores were not directly visible, but there were statistical arguments for their existence in the images. Just above the noise, the images seemed to contain many more pixels that moved in the "correct" direction around the Sun from one image to the next than pixels that moved in the opposite direction.

This analysis met with skepticism from the start, because the number of Halley-like comets in orbits beyond Neptune that would result from extrapolation of these results was much higher than the expected number. Once other teams began to analyze the images, they soon concluded that there were no significant indications of comets. The error analysis had been too optimistic. The uncertainty of the results exceeded the number of claimed objects by two orders of magnitude. Therefore, it became impossible to prove the existence of a population of Halley-sized Kuiper belt objects from these data. It remains an open question whether it will be possible to obtain new WFPC2 images with longer exposure times, or images with the Advanced Camera for Surveys after 2000, that might provide more information about these elusive objects.

Neptune and Uranus

Neptune

Although Hubble's observations of Neptune were confusing at first, a pattern seemed to emerge after 1994. Every few years, this distant gaseous planet appears to readjust the flow in its atmosphere, which then remains stable. This stability is expressed in the pattern of low-contrast bands on Neptune; at the same time, large atmospheric vortexes come and go. When the *Voyager 2* space probe visited Neptune in 1989, the most prominent feature was the Great Dark Spot (also called GDS-89), which had a bright companion high in the atmosphere that was also barely discernible in ground-based telescopes. Between 1990 and 1992, the dark spot had disappeared. During 1993, Neptune's northern hemisphere began to show a number of new white clouds, which still existed when Hubble obtained the first images with its corrected optics. By then, there were clouds from the pole to the equator, and a new

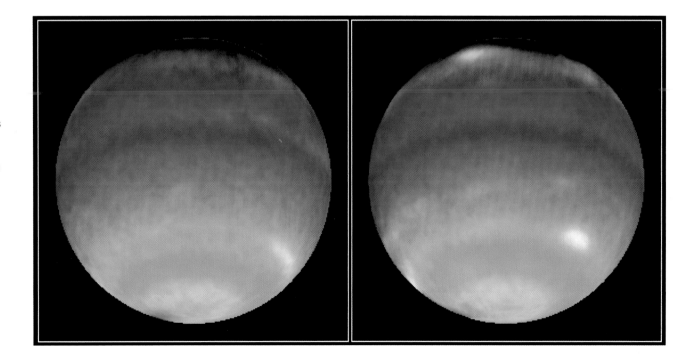

The weather on Neptune, made visible by clever selection of color filters. These two images, taken on August 13, 1996, show half a revolution of the distant gaseous planet. The basic blue color corresponds roughly to the true color of the planet; the colors of the clouds, however, encode their approximate altitude: the redder, the higher. (Sromovsky and NASA).

Great Dark Spot (GDS-94) had developed near the north pole.

The new dark spot, like the previous one, was accompanied by white clouds high above it. These large vortexes appear to influence the atmosphere over a large altitude range and to initiate the formation of clouds, just as mountains do on Earth and Mars. These observations clarified the behavior of Neptune over the past decades. Bright white spots had been observed on the surface of the planet since 1948. Because both Great Dark Spots had been accompanied by white clouds that disappeared when the vortexes subsided, it could be assumed that the white spots of the previous decades were visible indications of dark spots that could not be observed at the time: the dark vortexes themselves always had too little contrast to be seen in ground-based telescopes. But the Hubble Space Telescope has been able to see these details of Neptune's meteorology clearly since 1994. It also turned out that the white clouds rotate around Neptune at the same rate at all latitudes; they do not exhibit the differential rotation of the lower atmosphere that *Voyager* had measured. Rather, they rotate with the same speed as the polar regions.

This rotation probably indicates that the white clouds are carried by some kind of wave that is triggered by the dark spots. The sudden appearance of a peculiar oblique white cloud, covering 20 degrees in latitude, supports this theory. Such a cloud could not have been generated by any known type of circulation. Between 1995 and 1996, Neptune changed its appearance again. The large white clouds in the north had disappeared, but the northern dark spot was still there. In 1996, the dark spot had disappeared as well, but two southern bands of clouds remained. A detailed comparison with the *Voyager* images showed that these two bands had not existed in 1989. When new HST images were taken on August 13, 1996, a new white cloud—again connected to a dark spot— had appeared at extreme northern latitudes. It is very likely that this one is different from the one seen in 1994. Hubble will continue to inspect Neptune's clouds and bands in the future.

Uranus

The atmosphere of Uranus may not look as exciting as Neptune's, possibly because Uranus does not possess a strong interior heat source, as Neptune does. Nonetheless, Hubble did provide views into the layers of the atmosphere of Uranus, thanks to the WFPC2 camera and the appropriate selection of color filters; those in the near infrared provided particularly useful information. The NICMOS camera also photographed Uranus. Theses images, taken at wavelengths between 1.1 and 1.9 microns, differ only in small details from those observed earlier in visible light. The light from the Sun is reflected in the atmosphere of the planet, but at altitudes that differ depending on the wavelength used. The clouds at the right edge, which appear with more contrast in the infrared than in the visual range, are as large as entire

Clouds, rings, and moons of Uranus are seen in a false color infrared image composed of two NICMOS exposures taken 90 minutes apart on July 28, 1997. Three colors—from wavelengths of 1.1 to 1.9 microns—were combined; the resulting contrast makes six clouds visible. Each individual cloud is almost as large as the European continent. The rings can also be seen better in the infrared than in the visible. The moons—with errors indicating their movement over 90 minutes—have been artificially enhanced. (Karkoschka and NASA)

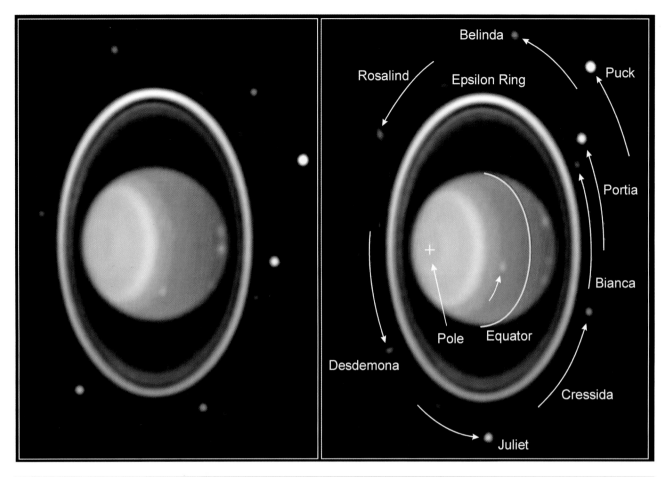

WFPC2 has also observed the clouds of Uranus. In the left image, taken at a wavelength of 547 nanometers, the colors were chosen to mimic the approximate colors that would be seen by the human eye. In the right image—taken at 619 nanometers, where methane in the planet's atmosphere absorbs sunlight—the cloud structure becomes more obvious. (Hammel and NASA)

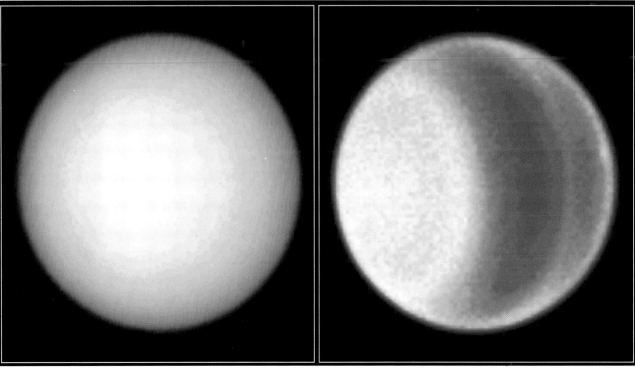

continents on Earth. The rings of Uranus, and 8 of the 10 small moons discovered by *Voyager 2*, are also clearly visible. The varying thickness of the brightest ring (Epsilon) is quite obvious. The smallest moons had not been seen since 1986; their diameters range from Bianca's 40 kilometers to Puck's 150 kilometers. All the inner moons of Uranus revolve around the planet in less than a day.

Saturn

The spectacular ring system of the giant planet Saturn has been intensively investigated by various means, including the Hubble Space Telescope. In 1995, Saturn and Earth were aligned in such a way that we were able to see the rings exactly edge-on—a rarity. The bright but very thin rings became almost invisible, which allowed us to see a number of other phenomena—such as extensions of the ring system on the outside and small moons—that normally would be drowned out by the bright light. Both Hubble and large telescopes on the ground observed a number of apparent moons, but closer inspection usually revealed that they were only temporary concentrations in the rings. Many of these "moons" had shapes that were too elongated, or they varied suspiciously in brightness. In addition, the *Voyager* probes should have found all moons of the corresponding sizes when they traversed the Saturn system in 1980 and 1981. Only a single new discovery turned out to be a previously overlooked new moon.

All the other phenomena in the Hubble images seem to have been small clouds of debris. Several mini-moons too small to have been detected by the *Voyager* probes may have collided with other small fragments of material in the rings and shattered into pieces. This would have increased dramatically the total surface area that reflects sunlight and allowed the collection of fragments to become visible. Such a phenomenon was no great surprise; it has been known for some time that Saturn's rings are a very dynamic system, the material in the rings being constantly replenished with numerous fragments of shattered small moons. This scenario is also supported by the fact that most of the small clouds of debris discovered in 1995 were located near the F ring of Saturn. This ring forms the outer edge of Saturn's main rings and constitutes the transition between these rings and the region of the larger moons. Its structure has been analyzed in detail since the edge-on images were obtained, revealing about 30 different condensations at least 20 kilometers in diameter, all of them temporary.

Observations made during the edge-on position of the rings also showed mysterious deviations of some moons from their predicted positions. The *Voyager* images of 1981 should have provided much more accurate positions. The moon Atlas, for instance, was identified in Hubble images only with great effort; it was found 27 degrees from the expected location in its orbit. Equally dramatic was the case of Prometheus, which lagged behind in its orbit by 19 degrees. This could be a result of interactions with the F ring, which it crosses every 19 years, or, more likely, of a close encounter with another moon. A candidate for such an encounter, however, was not found.

Another Hubble discovery during the edge-on observations was the unexpected curvature or warp of the entire ring system. This curvature became obvious when the transition of Earth (including HST) through the plane of the ring system during August 1995 occurred at slightly different times for the parts

A colorful infrared image of the planet Saturn, observed with NICMOS in January 1998 and presented at the eighth anniversary of Hubble's launch in April 1998. Exposures at wavelengths of 1.0, 1.8, and 2.1 microns were associated with the colors blue, green, and red, respectively, making the cloud structures more visible. Saturn generally appears to be yellowish-brown in color images taken in visible light. Chemical or physical differences in the cloud structures in several regions of Saturn's atmosphere are more pronounced at longer wavelengths. Even the rings of ice particles of various sizes appear more colorful here. (Karkoschka and NASA)

of the ring system to the left and to the right of Saturn. The gravitational influence of Saturn's moons was insufficient to explain the observed warp. Rather, it appears that the strangely irregular F ring is the cause. In Hubble images taken in November 1995, when the Sun illuminated the rings exactly from the edge, the three-dimensional structure of the F ring and its deviation from the plane of the other rings was revealed. Unfortunately, this effect prevented direct measurement of the thickness of the main rings during the edge-on position.

Hubble images of the inconspicuous E ring taken in August 1995 made it clear that this ring was brightest in the vicinity of Saturn's moon Enceladus. This observation proves that Enceladus is the source of small particles that replenish the E ring. The thickness of the E ring increases toward its outer edge; it is blue, and it consists of particles with a size of about 1 micron.

All these effects could be observed only because of the favorable geometric constellation in 1995, when the ring system was seen edge-on. But Hubble also provided information about other dynamic processes on Saturn: water from the rings "rains" down onto the planet! One of Hubble's spectrographs pro-

vided evidence of ultraviolet absorption features from water in Saturn's atmosphere. Moreover, there is a clear correlation between the location of water in the atmosphere and the position of Saturn's magnetic field relative to the rings. Water molecules knocked from the ice particles in the rings receive a negative charge and then travel along magnetic field lines into the atmosphere. Twice as much atmospheric water is found in places from which magnetic field lines lead to a ring as in places where the field lines point to a gap between the rings. The derivation of an exact erosion rate for the rings still requires major modeling efforts.

Another interesting Hubble observation was made in conjunction with the "raining" rings. The Faint Object Spectrograph discovered a thin hydroxyl (OH) atmosphere *around the rings* during the edge-on position on August 10, 1995. From their water ice particles, the rings produce between 10^{25} and 10^{29} OH molecules per second, which form a thin cloud around the rings. When the *Cassini* probe arrives at Saturn in 2004, it will find processes on Saturn and in its magnificent rings more complex than had been dreamed of during its construction.

Left: Retrospective of the edge-on position of Saturn's rings in 1995. The image on top shows the unlit side of the rings in November 1995, when the Sun was located south, and Earth north, of the plane of the rings. In this rare constellation, structures that are normally dark appear bright, and vice versa. From the outside, the F ring, the Cassini division, and the C ring are visible. During August 1995, Hubble viewed the rings almost exactly edge-on, as well as the large moon Titan (to the left of the planet), whose shadow falls onto Saturn's surface. Additional moons of Saturn are also visible. (Nicholson, Karkoschka, and NASA)

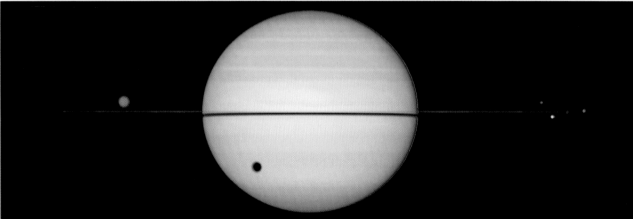

Right: The aurorae of Saturn as never seen before. With its high ultraviolet sensitivity, STIS was ideal to capture the glow in the upper atmosphere. At these wavelengths, it is more sensitive than any other Hubble instrument. The emission, induced by charged particles in Saturn's magnetic field, reaches an altitude of 2000 kilometers above the clouds. (Trauger and NASA).

Jupiter's Aurorae

Planets with magnetic fields are found to display luminous curtains similar to the aurorae on Earth. This phenomenon occurs when charged particles traveling along the planet's magnetic field lines collide with gases in the upper atmosphere, stimulating them to fluoresce or glow. The aurorae of the giant planets emit most of their light in the ultraviolet part of the spectrum. The best Hubble observations of these aurorae come from the new STIS instrument. Although it is mainly a spectrograph, STIS also serves as Hubble's best ultraviolet camera and can provide images of the aurorae even in the planet's sunlit hemisphere. STIS is about 10 times more sensitive in the ultraviolet than other Hubble cameras, which allows shorter exposure times. The instrument also provides two to five times better resolution in this range than WFPC2 or FOC. The aurorae several hundred kilometers above the clouds of Jupiter have now become clearly visible. The aurorae on Earth appear very similar when viewed from the space shuttle.

Jupiter displays another special effect: the "footprint" of the Io flux tube. Jupiter's strong magnetic field provides a powerful bond between the planet and its strange moon Io, whose volcanoes constantly catapult matter into surrounding space. Currents of millions of amps flow between Io and Jupiter. Where this flux tube interacts with Jupiter, various luminous effects appear in several spectral regions. The luminous processes in the ultraviolet, caused by the energetic particles of the flux tube in Jupiter's upper atmosphere, last for several hours, as the STIS images

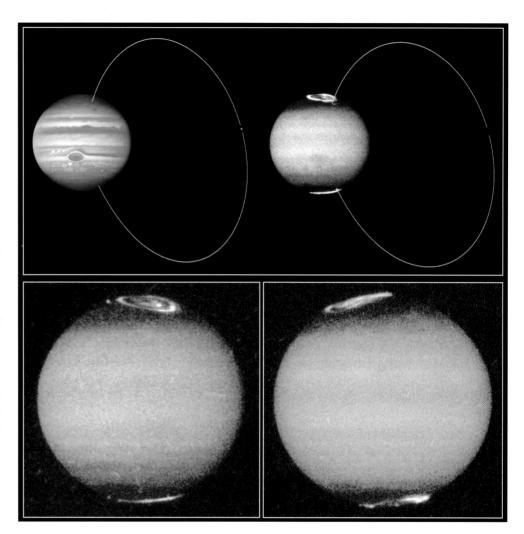

demonstrated for the first time. Only the new instruments on Hubble were able to show this effect; not even the *Galileo* probe was able to see them, since it lacked an ultraviolet camera. During October 1997, STIS also delivered images of aurorae on Saturn with previously unseen details.

Hubble observed the planet Jupiter, with its spectacular bands and swirls of clouds, in support of the *Galileo* mission. Because of a defect in its main antenna, *Galileo* could not transmit as many series of images of Jupiter as originally planned. Clever selection of targets allowed *Galileo* to remain a very

Left: The connection of Jupiter's aurorae with its moon Io. A so-called flux tube connects the volcanically active moon to the clouds of the planet. As particles move along the flux tube and hit the clouds, luminous phenomena appear in the ultraviolet. Closer to the poles, ovals of aurorae emerge (top right). They are caused by particles in Jupiter's magnetosphere. In visible light, the aurorae are too faint to compete with the sunlight (top left). The aspect of the aurorae changes with the rotation of the planet (bottom). (Clarke et al. and NASA)

Right: The sharpest images of Jupiter's aurorae were again taken by Hubble's STIS— and they show a level of detail that has never been seen before, not even by planetary probes near Jupiter. Particularly noteworthy are the footprints of Io next to the ovals of polar aurorae, where the flux tube from the moon connects to the planet. Because Io circles the planet more slowly than Jupiter rotates, the footprints move along the surface, leaving a glowing trail that diminishes slowly. The STIS images of September 20, 1997, were superimposed on a WFPC2 image taken the same day. (Clarke and NASA)

Jupiter looks quite different in the infrared from how it looks in visible and ultraviolet light. Three NICMOS images cover part of the planet. The images were taken at the wavelength where methane absorbs sunlight, which emphasizes the highest cloud layers. Also visible are the thin ring of the planet, seen from the side, and the moon Metis close to its edge. (Beebe and NASA)

fruitful mission, but the spacecraft could not monitor Jupiter's atmosphere continuously. Hubble was able to help by providing images of Jupiter's clouds to the *Galileo* controllers. To follow certain features on the planet, they had to know their location in Jupiter's turbulent atmosphere. The place where the *Galileo* landing probe entered the atmosphere in December 1995 was also monitored with the Hubble Space Telescope. These images helped scientists to interpret the surprising measurements returned by the probe, which had unexpectedly entered a particularly dry region.

Jupiter's Moons

Io

In addition to its luminous aurorae in the ultraviolet and its complex cloud structure, Jupiter also has a remarkable system of moons. The four Galilean moons are particularly large and complex. Which of these worlds may be the most interesting is a matter of a good-natured controversy among *Galileo* scientists. But for observers from the ground or from Earth's orbit, the answer is clear: Io. No other moon undergoes such rapid and dramatic changes, which are visible even over a great distance. Infrared telescopes often witness a temporary brightening of Io caused by yet another of its many volcanic eruptions. Hubble observed the results of such an eruption in July 1995: suddenly, a bright yellow spot about 300 kilometers in diameter appeared on Io's surface. Nothing conspicuous had been seen there a year earlier, and nothing like it had been observed during the previous 15 years. The position of the spot coincided with the location of the volcano Ra Patera, known from earlier *Voyager* images.

Hubble would soon observe another of Io's features. The large plumes from volcanic eruptions on Io had been seen only from space probes in the Jupiter system, first the two *Voyager* spacecraft and then *Galileo*. But this changed after Hubble took an image of Io in front of Jupiter's disk on July 24, 1996. The image, with a resolution of about 150 kilometers, was taken in support of the *Galileo* mission. Although *Galileo* can deliver images with more de-

tail at times, Hubble has the advantage of constant observational conditions. It also has ultraviolet filters, which are missing on *Galileo*. A year later, the details in the Hubble image were discovered: a 400-kilometer-high volcanic plume from an eruption of Pele could be seen. This image provided new impetus for monitoring Io's volcanic activity from a great distance.

Io displays yet another interesting phenomenon in the ultraviolet, as seen in an unusual exposure taken with STIS. The new Hubble instrument showed glowing hydrogen gas over the poles of Io for the first time. The origin of this hydrogen and the mechanism that makes it glow are both unclear, but the most likely sources are frost layers on Io's poles that contain hydrogen—hydrogen sulfide in particular. Io may have received the hydrogen via the magnetic flux tube; this connection between Jupiter and Io could have transported atoms from the planet's atmosphere to the moon. This phenomenon does not seem to be connected with the glowing belts of oxygen over Io's equatorial regions, however. Hubble's discoveries have shown us that the strange world of Io is an even more complex place than we thought.

Ganymede

The Hubble Space Telescope can resolve details on any of the four largest moons of Jupiter, but sometimes it is more interesting to observe their spectra. For instance, Hubble found that Ganymede

A volcanic eruption on Io, in front of the disk of Jupiter. This image in violet light was taken on July 24, 1996, in support of the *Galileo* space probe, which had been orbiting Jupiter since the end of 1995. Several months later, a more detailed analysis of this image and an ultraviolet image taken around the same time showed that Hubble had accidentally witnessed a volcanic plume on Io (image at bottom, composed of both colors). Only planetary probes near Io had been capable of such a view up to this time—and a plume reaching an altitude of 400 kilometers had never been seen. The volcano that causes these eruptions, Pele, has been known for quite some time. (Spencer and NASA)

possesses an aurora that was too faint to be seen by *Galileo* during its rapid passes. By taking a spectrum in the ultraviolet with the Goddard High-Resolution Spectrograph over five full orbits, Hubble detected a faint glow over both poles of this moon. Although Ganymede's aurora is much fainter than Jupiter's, or even Earth's, its physics are the same. The magnetic field lines of celestial bodies that possess a magnetic field divert electrons and other charged particles toward their poles, where they collide and excite gas molecules in the atmosphere, causing the oxygen molecules to radiate ultraviolet light. Because the *Galileo* probe had found repeated indications of a magnetic field and an atmosphere on Ganymede, the faint aurora did not come as a great surprise.

The oxygen in Ganymede's atmosphere, which is as dense as Earth's atmosphere at an altitude of several hundred kilometers, originates in the ice on the moon's surface. It is released from the ice by charged particles from Jupiter's magnetosphere, by photons from the Sun, and by micrometeorites. Even ozone is present, as Hubble's Faint Object Spectrograph discovered in 1995. Charged particles are trapped in Jupiter's strong magnetic field and are caught up in the rapid rotation of the planet. Because Ganymede orbits Jupiter at a much slower speed, these particles hit the moon "from behind" and enter its ice layers. There, they destroy water molecules and produce ozone in a process that can be replicated in lab experiments.

Asteroids

Vesta

Hubble has observed the numerous small planets between the orbits of Jupiter and Mars. The most detailed observations were carried out for the asteroid Vesta. Hubble provided systematic images in several colors of this 530-kilometer-long, potato-shaped object. The map resulting from these images shows that Vesta has a strongly variable surface, unlike most other asteroids. The various areas could be old lava fields or areas where impacts have removed the lava to expose the underlying material. The colors indicate that the entire surface consists of igneous rock. Either Vesta was entirely molten at some point in its past, or lava from the interior has flooded the entire surface. Even better Hubble images provided another surprising discovery in 1996: Vesta's surface contains a huge impact crater near its south pole. With a diameter of 460 kilometers, it is almost as large as the entire asteroid. The cosmic collision that produced this crater must have ejected about 1 percent of the asteroid's mass into space. We may even hold a few pieces of this event in our hands: about 6 percent of all meteorites that fall to Earth are probably material from Vesta.

Serendipitous Discoveries

Hubble's extremely narrow field of view makes the telescope inappropriate for science projects that require a systematic survey of large parts of the sky. Wide-field telescopes on the ground are much better suited for such programs. But the combination of many sets of archival Hubble data taken over long periods of time may take on the character of a sky survey. In this way, Hubble made a name for itself as an asteroid counter. In a project carried out mainly as an extended quality check of Hubble's most important camera, WFPC2, astronomers noted bright, curved lines in several images, which had nothing to do with the intended target. The lines were caused by asteroids moving along their orbits during the exposure, while Hubble raced around Earth at the same time. The combination of these two motions allowed the distance of the asteroids to be determined directly, a feat that is impossible from a single ground-based image. The analysis of more than 28,000 WFPC2 images in the Hubble Data Archive yielded about 100 asteroids, all of them in the main belt between Jupiter and Mars. Extrapolating from this analysis allows us to estimate that there are about 300,000 individual bodies with diameters of 1 to 3 kilometers between the orbits of these two planets.

Things that cannot be seen in these images are also of interest. The WFPC2 images did not show a single alleged mini-comet; if they existed, these images should have shown thousands of them in the immediate vicinity of Earth. The absence of these objects undermined the questionable hypothesis that Earth is constantly bombarded by snowballs with diameters of about a meter. Astronomers had never

Two maps of the asteroid
Vesta, assembled from an
extensive series of Hubble
observations in 1995.
The asteroid surprisingly
shows a dark and a light
hemisphere (top). For the
dark area (perhaps a lava
flow?), the name Olbers has
been proposed, after the
astronomer who discovered
Vesta in 1807. The chemical
composition of the surface is
charted at the bottom. Again,
two different hemispheres
are apparent. (Zellner and
NASA)
Translation of German terms
    (English in parentheses):
Breite   (Latitude)
Helligkeit   (Brightness)
Chemische Zusammensetzung
    (Composition)
Länge   (Longitude)

Hubble was able to continue
its observations of the
asteroid Vesta in 1996. A
typical single image with
a resolution of about 10
kilometers is shown at
top left. A color-coded
altitude map (bottom)
was calculated from 78
individual images and
revealed a depression with a
diameter of 460 kilometers.
The computer-generated
composite of Hubble images
of Vesta (top right) shows
that the giant crater has a
central peak close to the
pole of the asteroid. (Zellner
& Thomas and NASA)
Translation of German terms
    (English in parentheses):
Modell   (Model)
Höhe   (Elevation)

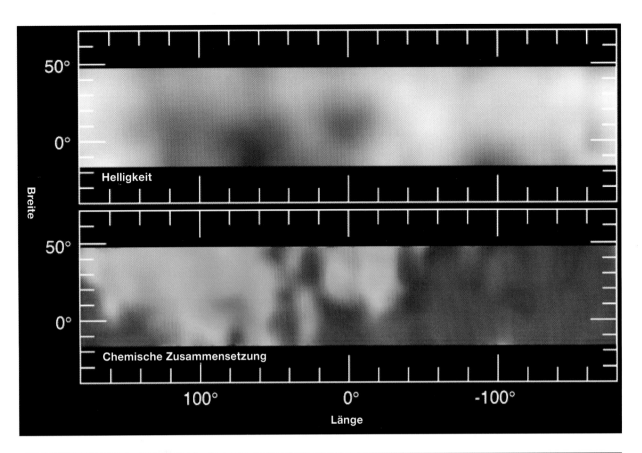

Serendipitous discoveries of asteroids from images in the giant Hubble Data Archive. The curved lines are the trails of asteroids that moved through WFPC2's field of view during exposures of other targets. All asteroids shown here are from 800 meters to 2 kilometers in size and at distances of 140 million to 400 million kilometers from Earth. Breaks in the asteroid trail at bottom right (in front of the outer regions of the galaxy NGC 4548) result from interruptions in the exposure caused by Earth moving between Hubble and its target. (Evans & Stapelfeldt and NASA)

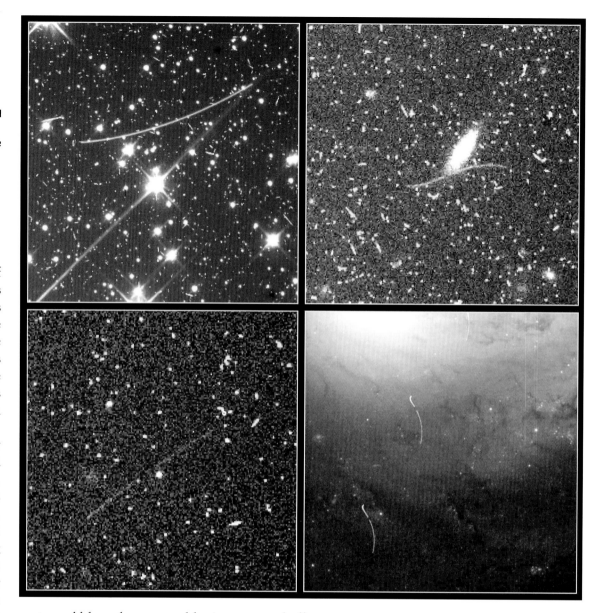

embraced this hypothesis, for a constant rain of comets was not required to explain any mysterious observations, nor did these objects have an obvious source. Some geophysicists, however, had seen a large number of short-lived dark spots in front of the bright atmosphere in their ultraviolet satellite images of Earth. Most image analysts believed that these spots were simply detector noise, but a few took this supposed phenomenon seriously and searched for an explanation.

The great comet bombardment was the only hypothesis that had not been ruled out, and it caused quite a stir among geophysicists on two occasions, in 1986 and in 1997. Both times, many arguments were quickly brought forth to refute the theory. Most damaging was the lack of results in the actual search for the purported mini-comets. In the spring of 1998, the analysis of the archival Hubble images and results from a systematic ground-based search for small planets reported the lack of a single observation of these mini-comets. The Space Watch Telescope at Kitt Peak National Observatory in Arizona should have found about 2000 of them during its seven-year search, yet it did not see a single one (although it did discover thousands of asteroids). In addition, studies appeared that showed how the original dark spots could have been caused by instrumental effects in the cameras themselves. This will probably mean the end of the great comet bombardment hypothesis. The Hubble Space Telescope played a completely unexpected role in its demise.

Comets

With its tiny field of view, Hubble can study only the cores of comets and their comas. One recent series of Hubble observations involved the comet Hale-Bopp, which could be seen easily by the naked eye during the spring of 1997. Hubble's first image of the comet was taken on September 26, 1995, only two months after its discovery, when Hale-Bopp was still far from the Sun. This remarkable image showed a spiral-shaped dust structure that had grown from the core of the comet a short time previously. Such a structure had appeared once before but had disintegrated rapidly. Luckily, the Hubble observation, which had been planned weeks in advance, captured the second event. At that time, astronomers could only speculate about this unusual phenomenon: the core of the comet appeared to rotate like a lawn sprinkler, expelling dust into space.

When Hubble took the next image on October 23, the coma of Hale-Bopp was completely featureless, with no trace of a dust fountain. The absence of the dust structure allowed astronomers to analyze the brightness distribution of the comet mathematically and to isolate the fraction of light coming from the solid core. They determined that Hale-Bopp was about 40 kilometers in diameter, comparatively large for a comet. Then Hubble had to take a break because Hale-Bopp was too close to the Sun to take further observations during its brightest phase. By the summer of 1997, Hale-Bopp could be observed again, and now the newly installed STIS could be used to take spectra. They showed emission of hydroxyl (OH), a molecule resulting from the decomposition of water, at a wavelength of 309 nanometers. It was estimated that the comet core released about 3×10^{29} water molecules per second in August 1997, about as much as it did at the same distance from the Sun on its way into the solar system.

The dust spiral of comet Hale-Bopp on September 26, 1995. By chance, a Hubble observation had been scheduled shortly after a strong eruption from the surface of the core of the comet; it was still quite distant from the Sun at this point. In the image at right, all traces of background stars have been removed for clarity. (Weaver & Feldman and NASA)

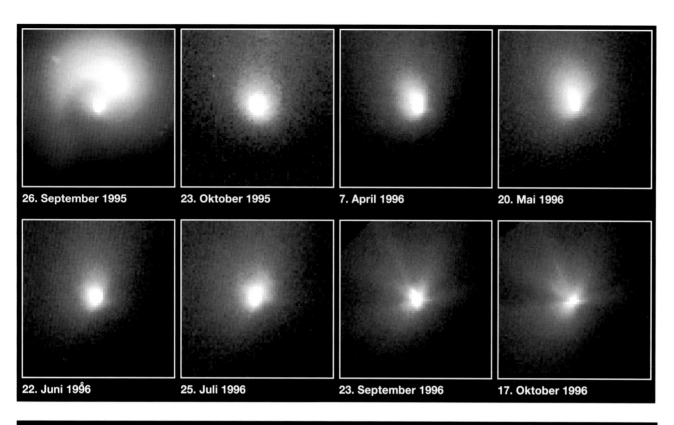

26. September 1995   23. Oktober 1995   7. April 1996   20. Mai 1996

22. Juni 1996   25. Juli 1996   23. September 1996   17. Oktober 1996

A year of Hubble observations of Hale-Bopp, from September 1995 to October 1996, documents the changes in the inner gas and dust layers of the comet. Although occasional dust eruptions dominated at first, a stable pattern of dust rays appeared after the fall of 1996. (Weaver and NASA)

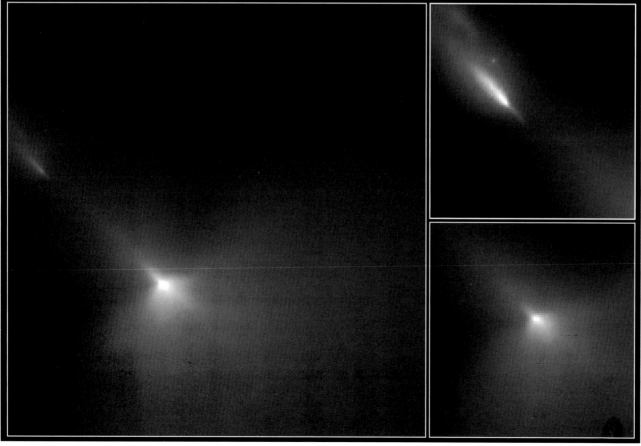

The innermost parts of a comet are shown in Hubble images of comet Hyakutake on March 25, 1996, when it was only 15 million kilometers from Earth. The left image covers an area of only 3340 kilometers. Most of the dust is released on the Sun-facing side of the comet core; because it is only a few kilometers in diameter, it is too small to be resolved even by Hubble. Several fragments have broken off from the core and move away from the Sun (toward top left). An enlargement of the fragments can be seen in the image on the right. (Weaver and NASA)

Mars

The Hubble Space Telescope has often observed Mars in support of other space missions. During the months before the arrival of the *Mars Pathfinder* mission at the red planet, scientists tried to provide accurate weather predictions from Hubble observations. By 1995, a significant change in the climate over the past two decades had become obvious. *Pathfinder* was expected to see a Martian sky quite different from the one seen by the *Viking* lander in 1976. White clouds on a dark blue Martian sky were anticipated for *Pathfinder*'s landing on Mars on July 4, 1997, as forecast by NASA on May 20, based on the latest Hubble images. In contrast, the *Viking* probes had seen a yellow sky. Hubble observations during March 1997 had confirmed two basic climates on Mars: a warm one with a dust-filled atmosphere and

Mars on March 10, 1997, shortly before its closest approach to Earth. Even at a distance of 100 million kilometers, the Hubble images display a wealth of details. Observations were carried out using nine color filters to emphasize both surface features and cloud structures; the three resulting composite images show practically the entire planet. This observing campaign was part of the support for the *Mars Pathfinder* and *Mars Global Surveyor* missions. Both space probes were approaching the planet at this time. (Crisp & WFPC2 Team and NASA)

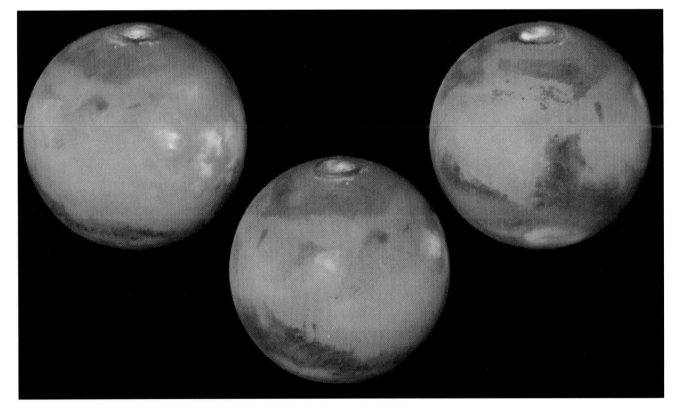

a cold one in which clouds of water vapor emerged.

Three factors seemed to influence the change from one climatic state to the other: the low density of the Martian atmosphere, the variable amount of sunlight, and the strong interaction between dust and clouds of water ice in the atmosphere. Because Mars does not have oceans to serve as heat reservoirs and its atmosphere does not extend very high, the planet's temperature reacts quickly and strongly to changes on the surface and to the heating of the atmosphere by the Sun. As Mars approaches perihelion (the closest point to the Sun in its orbit), the temperature in the southern summer can rise by as much as 20°C, triggering global dust storms that can increase the atmospheric temperature by another 15°C to 30°C. This phenomenon had been seen during the *Mariner 9* and *Viking* visits. The typical climate of Mars at aphelion (the most distant point from the Sun in its orbit) is quite different—the atmosphere is much cooler and clearer, and there are planetwide belts of water ice clouds at an altitude of 3 to 10 kilometers. A very dark and perhaps slightly blue sky with brilliant white clouds was therefore predicted for the July 4 landing. As with weather forecasts on Earth, however, this one turned out not to be accurate.

Pathfinder found a Martian climate only slightly different from the one seen by the *Viking* spacecraft. The sky was as yellow as ever, with a comparatively large dust content. What had gone wrong? Scientists could not find a simple explanation to account for the

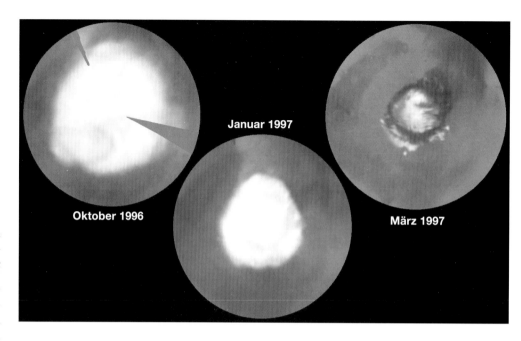

differences between the predictions based on Hubble's distant images and the conditions *Pathfinder* actually encountered. Apparently, it is more difficult than expected to distinguish atmospheric phenomena from details on the surface, even in the excellent Hubble images. But the space telescope still served as an important "weather satellite" in providing warnings of emerging dust storms. Hubble had observed such a dust storm in detail: a small, light brown cloud appeared over the north polar cap and developed into a major storm system over the entire cap a month later. This process had never been observed before. The unusual storm may have been caused by the large temperature differences between the polar ice caps and the more southerly regions warmed by the spring Sun.

This storm was too small to have been observed with ground-based telescopes; Hubble is the only instrument that can discern such local storms. Hubble's weather reports were important for both the *Mars Global Surveyor* and *Pathfinder*—their fate required accurate knowledge of the state of the Martian atmosphere, which was used to decelerate the spacecraft.

The north polar cap of Mars vanishes. Hubble observed the dramatic reduction in size of the planet's carbon dioxide ice cap with the increasing Sun angle between October 1996 and March 1997. Image sequences were projected onto views from above the planet's pole. The polar cap at first reached down to 60 degrees northern latitude, then assumed a roughly hexagonal shape, and finally almost disappeared except for a small patch of water ice. Before the Hubble era, images of this quality had been delivered only by planetary probes near Mars. (James et al. and NASA)

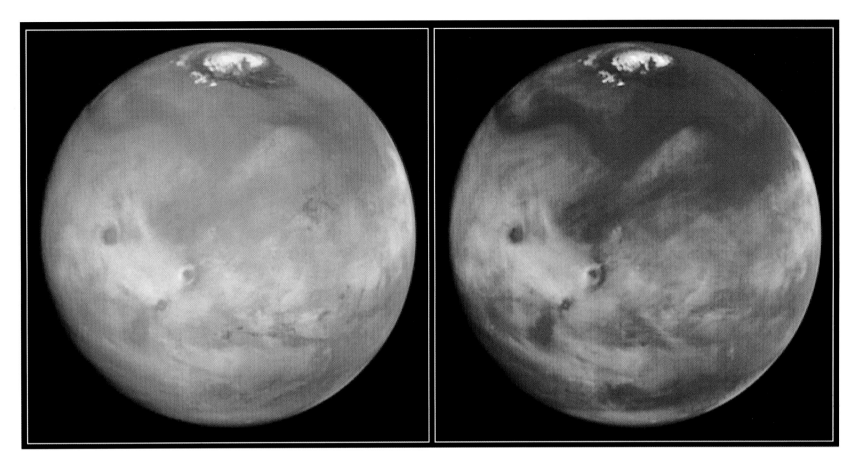

The various faces of Mars depend on the color filters. The composite at left results from red, green, and blue images, giving roughly the correct, or natural, color distribution. The blue image shown separately at right emphasizes the clouds on Mars, resembling images from terrestrial weather satellites. An extended frontal system appears, and global meteorological effects become visible. (James et al. and NASA)

Hubble images taken on June 27, 1997, only a week before the *Pathfinder* landing, caused some excitement: a new dust storm had formed 1000 kilometers south of the landing site. Further images showed that the storm had not approached any closer, however, and the landing proceeded as planned. Now that the *Mars Global Surveyor* is in orbit and will deliver daily weather maps beginning in March 1999, Hubble's importance for Mars science will decline. This is one of the very few areas of space science, however, for which the Hubble Space Telescope can no longer provide pioneering contributions.

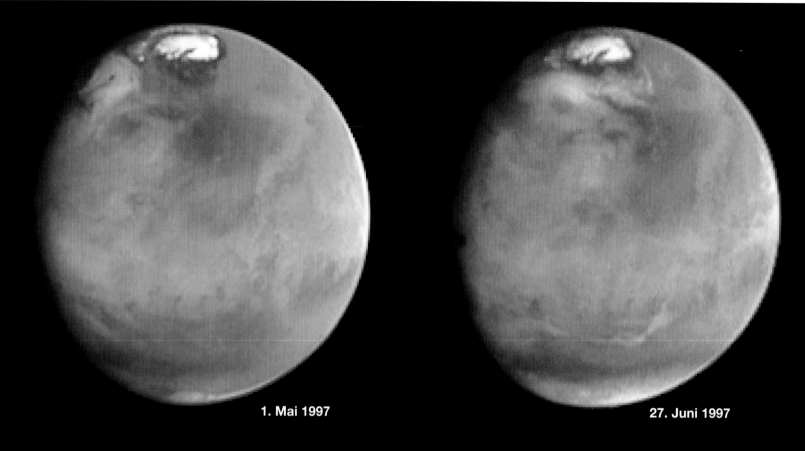

1. Mai 1997   27. Juni 1997

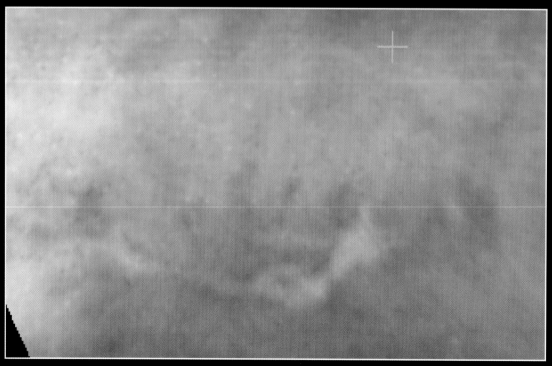

Valles Marineris   27. Juni 1997

Weather report for the landing of *Mars Pathfinder*. During the weeks before the first landing on Mars in 21 years, Hubble was often trained on the red planet. It discovered a dust cloud 1000 kilometers south of the landing site at the end of June 1997. When *Pathfinder* arrived on July 4, the storm had largely subsided, but new dust activity had appeared farther north. The top images show a global view of Mars; the *Pathfinder* landing site is marked with green crosses in the detailed views on the bottom. (Lee et al. and NASA)

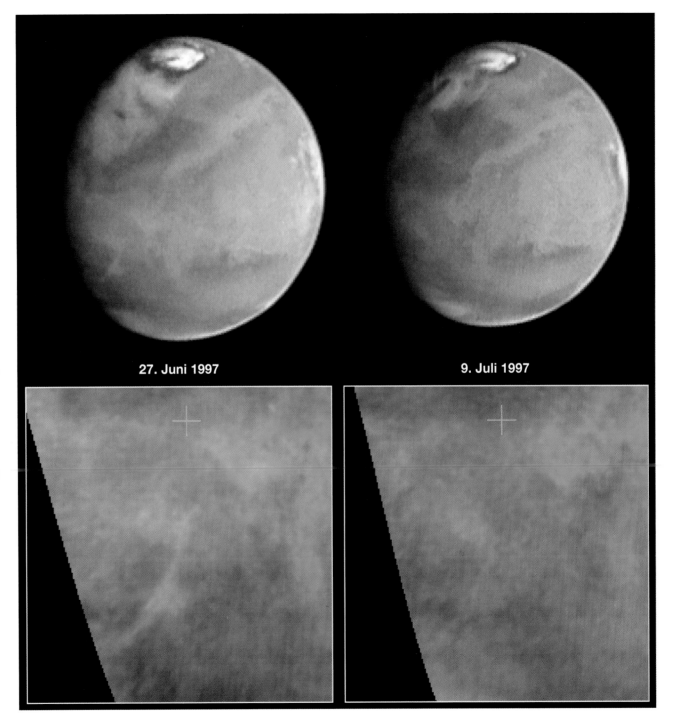

27. Juni 1997   9. Juli 1997

Part 5

Hubble's Future and Its Successors

Hubble's Second Decade

As a satellite, the Hubble Space Telescope usually does not attract much attention, except during visits by the space shuttle every three years. Rather, it fascinates by providing spectacular astronomical images and results. But on the occasion of a launch anniversary—most recently on April 24, 1998—it is time to take stock. In its first eight years, the satellite had traveled 1.9 billion kilometers in its low Earth orbit, more than the distance to Saturn. During that time, it transmitted 4.44 trillion bytes of data back to Earth, filling 710 large optical disks, each with a capacity of 6.7 gigabytes, in the central Hubble Data Archive. Only part of this wealth of data contains actual astronomical observations, but the Hubble Space Telescope has already provided about 120,000 exposures of approximately 10,000 different objects, leading to 1700 scientific publications. Who would have thought during the crisis shortly after launch that Hubble would turn out to be so productive? Even the optimists didn't expect 100,000 exposures until the 10th year of the mission, but this milestone was reached by June 1996, almost four years earlier than anticipated at launch.

One reason for the increased yield of data is that Hubble does not need as much overhead as planners had thought. Original estimates assumed that only 35 percent of the time would be spent on astronomical targets; almost half of each typical orbit, after all, is lost simply because the target disappears behind Earth. Other losses result from the Earth's radiation belts, particularly the South Atlantic Anomaly, which

demands a pause in the observations because of the high particle background. But increasingly clever planning and scheduling of observations, and more effective command of the scientific instruments, increased the observing efficiency to 38 percent in 1992, 47 percent in 1995, 52 percent in the first quarter of 1996, and 55 percent by mid-1996. The scientific output is now almost twice as high as originally hoped for. The efficiency record—an unbelievable 74 percent—was set in December 1997 when Hubble observed the identical field in the sky for almost 10 days: the famous Hubble Deep Field. Because the target was located close to the pole of HST's orbit, it was not obscured by Earth during the satellite's revolutions around it; that had been one of the main selection criteria for the field.

As we have seen, Hubble has provided important contributions to practically all areas of astronomy, and its use in answering central questions has become almost a matter of course. Although it offers unique possibilities, the Hubble Space Telescope is an observatory like many others. But observatories on the ground have an unlimited life span, at least in principle: several institutions are more than a century old, and even the famous 5-meter telescope on Palomar Mountain celebrated its 50th anniversary in 1998. Satellites in orbit have a limited life span— and the halfway mark of Hubble's original mission duration of 15 years was passed at the end of 1997. Questions about the future of optical astronomy in space become more pressing, but a clearer vision of the future now begins to emerge. For the foreseeable future, NASA will maintain its near-monopoly in orbiting observatories for the ultraviolet, visible, and infrared regions of the spectrum. A new and creative approach will develop projects that require only a fraction of the cost of the Hubble program. It also appears likely that international collaboration, particularly with the European Space Agency, will continue in the future.

NASA made an important decision in 1997: instead of terminating the Hubble mission in 2005, NASA will continue to operate the telescope until at least 2010, provided that the various systems continue to work and that the scientific output remains high. One of the prerequisites for extending the mission was the decision to leave Hubble on its own after the fourth servicing mission, now planned for early 2003. By not planning any further visits by the space shuttle, NASA will significantly reduce annual operating costs and avoid the costs of development of new instruments and spare parts. These resources can be devoted to the development of Hubble's planned successor, the Next Generation Space Telescope (NGST). If this new facility is launched in 2007, as planned, it will work in parallel with Hubble for several years. The Hubble Space Telescope of this era will be radically different from the one launched in 1990: none of the original instruments will be on board, and many of the mechanical and electronic components, including the solar panels, will have been exchanged.

A New Director for the Space Telescope Science Institute

The second change in the leadership of the Space Telescope Science Institute (STScI) in Baltimore, Maryland, occurred on September 1, 1998: Steven V. W. Beckwith became the third director of STScI, replacing Robert E. Williams, who had announced his wish to step down some time previously. Williams had not grown tired of HST — after all, the two servicing missions and some of Hubble's greatest triumphs occurred under his watch — but the astrophysicist and former director of the Cerro Tololo Interamerican Observatory in Chile longed for more time to do active science and resume his scientific work.

Beckwith comes to STScI from the Max Planck Institute for Astronomy in Heidelberg, Germany, where he has been managing director since 1994 and a director since 1991. He earned

Steven V. W. Beckwith, the new director of the Space Telescope Science Institute. (MPIA)

187

his Ph.D. from Caltech in 1978 and then moved to Cornell University for the first part of his career, becoming Professor of Astronomy in 1989. Among other distinctions, he was an Alfred P. Sloan Fellow from 1982 to 1985.

He is well known as an infrared astronomer and developer of instrumentation. As a graduate student at Caltech, he designed and constructed a thermal infrared photometer that he used at Mount Wilson. His thesis dealt with molecular hydrogen emission from many different sources, such as planetary nebulae, young stars, and interstellar matter. He continued this line of work when he went to Cornell and built a new spectrometer for the Kuiper Airborne Observatory to study molecular lines and broad dust features in the thermal infrared. He also began work on stars with protoplanetary disks. The study of these disks and their properties has been fundamental in estimating the number of planetary systems in our galaxy. He also acquired an interest in early galaxy formation while at Cornell and started a project to observe the infrared signatures of primeval galaxies, continuing in this area after he went to Heidelberg.

Upon his arrival at Heidelberg, Beckwith's goals were to equip the institute's observatory with modern instrumentation by decreasing the time and cost needed for instrument projects, to establish collaborations with other institutes, and to diversify the staff of about 190 individuals. These changes led to a faster pace of scientific output by a staff acting with greater self-direction. He has also been involved with HST in several key ways, including memberships or chairmanships of several review panels and advisory committees.

Beckwith considers STScI "one of the most exciting new institutes in astronomy" and expresses his pleasure about the appointment to the institute's directorship. "I hope that HST is only the first of many important space projects for STScI. I am particularly interested in having STScI actively participate in the NASA Origins Program." The institute will play an important role during the development of the Next Generation Space Telescope and may become one of the main centers for Origins missions. In addition to its primary technical and scientific tasks, STScI will continue to provide information to the general public about scientific results from the Hubble Space Telescope and other future missions. Its Office of Public Outreach is recognized for its achievements throughout the world.

The Next Two Shuttle Visits

The third servicing mission, now planned for spring 2000, will include six space walks of six hours each and will be at least as complex as the previous two missions in 1993 and 1997. No fewer than 10 different units or subsystems will be exchanged, the satellite's insulation will be repaired, and the altitude of its orbit will be increased significantly. Only one change is planned for the complement of scientific instruments: the European Faint Object Camera (FOC) will be removed to make room for the Advanced Camera for Surveys (ACS), which will become the most important Hubble camera. Its main advantage comes from its combination of a larger field of view (more than 3 arc minutes) and identical or better angular resolution. Additional improvements include better sensitivity and reduced stray light. ACS actually comprises three different cameras: a Wide-Field Camera with 16 million pixels, a High-Resolution Camera with 1 million pixels (both using CCD chips, similar to WFPC2), and a Solar-Blind Channel for ultraviolet light (using a technology related to that used in the FOC).

Earlier instruments on Hubble were predominantly used for the observation of individual objects, but the ACS will also be used for surveys. Certain regions of the sky will be observed systematically. First, images of an area of roughly two-thirds of a square degree will be taken with the Wide-Field Camera in two colors; about 20 large galaxy clusters will be observed in this manner. Then, the distances of galaxies will be determined independent of their redshift, to provide information about their large-scale motion and about the cosmological distance scale. Aside from these cosmological projects, ACS will conduct high-resolution investigations of the nuclei of active galaxies and quasars. Additional areas of research will include the aurorae of planets in our solar system and protoplanetary disks around other stars.

Another important task of the third servicing mission will be to replace the second generation of solar panels with smaller, less fragile panels. These new, rigid panels are similar to those used for the Iridium series of telephone communications satellites. Several internal components are also scheduled for exchange. Among them are a new Fine Guidance Sensor, an attitude control unit, one of the on-board computers (which now contains a 486 processor, as space-qualified hardware always lags behind the latest models on the ground), and two data recorders to complete the change from the older tape models to exclusively solid-state storage. At the time of this writing, the final decision about whether or not to install a new NICMOS Cooling System (NCS) had not been made; this unit will be tested under space conditions during a shuttle flight in late 1998. The NCS could provide continued life for NICMOS, whose original supply of cryogen will run out by the end of 1998. The continuation of repairs to the thermal insulation of the satellite and the boost to a higher orbit will also be important parts of the third servicing mission.

The fourth and last servicing mission in early 2003 will be fully geared to extending Hubble's life for the seven or more years that it will have to spend completely on its own (like many other research satellites). Several subsystems will be brought up to the latest standards to ensure their continued operation, and the last two new scientific instruments will be installed. The COSTAR corrective optics will be removed, since all its first-generation "customers" — FOC, FOS, and GHRS — will no longer be present. Its space will be taken by the new Cosmic Origins Spectrograph (COS), a low-cost unit that can supplement STIS in several areas. And the Wide-Field and Planetary Camera 2 will be replaced by the Wide-Field Camera 3 (WFC3). For the first time, the desire for redundancy plays a role here: if STIS or ACS were to fail during the later years of Hubble's mission, the new units could provide backup for the spectrograph and the main camera.

To ensure this redundancy and provide better backup for STIS, the capabilities of COS and its spectral range were increased from the original proposal. This was possible only because the original concept for COS was particularly economical. In its original form, COS was designed for special applications in the far ultraviolet, from 115 to 178 nanometers wavelength; but enough funds were made available to implement a second channel in the near ultraviolet, covering the range from 175 to 320 nanometers. The additional channel will make use of a flight spare for the STIS MAMA detector, again saving money. In addition, COS will be more sensitive over its more restricted wavelength range than STIS, providing higher sensitivity by factors of 20 to 40. COS will lack the one capability that makes STIS so effective, however: it will not be able to image an entire area of the sky but will be restricted to providing a spectrum of a single object at a time — albeit the faintest and most distant objects in the sky. The research projects foreseen for COS include the investigation of intergalactic and interstellar gas and the physical properties of planets, stars, and galaxies. Emission and absorption lines in the ultraviolet can provide significant insights into these objects. Only a very small part of the observing time is reserved for the team developing COS; the majority of time will go to general observers.

The replacement of WFPC2 by WFC3 will complete the change of generations. The new instrument, which will be inserted in one of the radial instrument bays, will look very much like the one it replaces. This is not surprising, because WFC3 uses the very box that housed the original WFPC and was brought back by the astronauts during the first servicing mission in 1993. But the interior shows the progress of technology: instead of the old, small detectors, a copy of the large CCD chip with 4096 by 4096 pixels, used for the Wide-Field Camera of ACS, will be installed. It provides WFC3 with a field of view of 160 by 160 arc seconds with a resolution of 0.04 arc second per pixel and can provide images as detailed as the Planetary Camera chip of WFPC2, but over a much larger area. The use of components from earlier Hubble instruments, such as WFPC, WFPC2, ACS, and STIS, provides additional cost savings and emphasizes WFC3's role as insurance against a failure of the main ACS camera in the later years of the mission.

Hubble's Legacy: The Data Archive

After the last servicing mission and the installation of the last generation of new instruments, Hubble's annual operating costs of more than $200 million will decrease dramatically: this new approach is called "Cheap Ops." In several areas, however, the amount of work will remain constant. Astronomers using Hubble will still require support—and the flood of data will have to be stored and made accessible to the astronomical community in convenient ways. Even if the data from a particular observation are proprietary at first, they generally become public after a year. The scientific community has recognized the value of this gold mine: usage of the several terabytes of data available in the archive is increasing constantly. In 1997, the main Hubble Data Archive at the Space Telescope Science Institute (STScI) in Baltimore, Maryland, experienced an average retrieval rate of 10 gigabytes per day; during February 1998, the rate increased to 25 gigabytes daily. Not surprisingly, the retrieval rate now exceeds several times the rate of incoming data.

The search for particular data sets or for material related to a specific topic has become increasingly simple, thanks to the development of new software; and the growth of the Internet and the appearance of writable CD-ROMs have made the distribution of data much easier. In addition to the main archive in Baltimore, which shoulders the majority of the Hubble data traffic, two more sites contain copies of the Hubble data: the Space Telescope European Coordinating Facility in Garching near Munich, Germany, and the Canadian Astronomy Data Centre of the Dominion Astrophysical Observatory in Victoria, British Columbia. The archive in Europe has already been converted from the expensive optical disks to CD-ROMs. Even though a CD-ROM has only a tenth the capacity of an optical disk, the change still provides cost and space savings. The Hubble Data Archive at STScI in Baltimore, however, still uses the optical disks, mainly because of throughput considerations and the major investment in hardware, specifically the huge jukeboxes that hold almost 100 optical disks each, ready for near-online access. But preparations have begun for the change to new storage technologies, because these systems become obsolete within only a few years.

The next step will be Digital Versatile Disks (DVDs). These disks, which look like CD-ROMs, will contain several gigabytes, compared to only 650 megabytes for the CDs. They were developed for the entertainment industry but can also serve as excellent data storage devices for computers. The software side of the archive will also change. Hubble data in their raw form cannot be directly used for science. Each observation has to undergo reformatting steps; more important, calibrations must be applied to remove the effects of the instruments and to convert the data into physically meaningful units. The volume of data increases dramatically in these steps, because several copies of raw, processed, and calibrated data sets have to be stored. In addition, the calibration parameters change, usually becoming much better over time. It turns out to be more efficient to store only the raw data and the calibration parameters and to perform all the computational reformatting and calibration steps at the time of retrieval.

This technique is called "on-the-fly calibration"; it will provide users with the best available version of the Hubble data, and it has become feasible as a result of the increased processing power of computers over the last few years. Pilot projects using this technique are already under way. In addition, ways to off-load the processing from the archive computers are being explored. An implementation of the calibration and analysis software in Java and similar languages would allow it to run on practically any user's computer with a Web browser. For the first time, the world of Hubble data could be available to virtually anybody with an interest in astronomical images and spectra, including amateur astronomers and the general public. In principle, the data are already accessible, but the sophisticated software systems needed to process them and make them visible are available for only a limited set of operating systems. However, the organizations running the data archives will not be able to provide comprehensive support for users other than professional astronomers.

The incredible wealth of data from the Hubble Space Telescope will probably guarantee the existence of the Hubble-related institutions and archives for quite some time, albeit on a smaller scale. But the satellite itself will cease to operate at some point in time. Once too many systems have failed or the demand for Hubble data has fallen below acceptable limits, it will have to be switched off—not, it is hoped, before 2010 or even later. But switching off the satellite is not sufficient. The noticeable drag by the atmosphere remaining at Hubble's altitude would lower its orbit, slowly at first and then increasingly faster, until it would enter the denser parts of the atmosphere and burn up. The risk of large parts of the telescope, particularly its mirror, reaching the surface is too high to allow that to happen. The Hubble Space Telescope does not possess thrusters of its own to lift it out of the decaying orbit. This leaves only two alternatives, both already being discussed within NASA. The first would use a robot satellite to lift Hubble into a higher and safer "graveyard" orbit. The second would use either a similar vehicle or the space shuttle to capture Hubble and bring it back to Earth. This second option would provide the engineers who developed Hubble 30 years ago, or their successors, the unique opportunity to study the long-term effects of space on the various parts of the satellite. Moreover, the Hubble Space Telescope could find its well-deserved final resting place in a museum.

Hubble's Successors

A world without a large optical telescope in space has become unthinkable for astronomers at the end of the 1990s—and they may not have to do without one. Since 1985, an increasingly clear vision of the future has been emerging. In pursuit of this vision, called "Origins," NASA is preparing an entire series of observatories in space to explore the first galaxies as well as Earthlike planets around other stars in ever increasing detail. The basic premise is clear: Where do *we* come from, and are there other civilizations out there? The Origins Program combines political and scientific considerations, for NASA tried to find cosmic questions of deep interest to the public, while at the same time seeking broad support from the science community. The answers to the grand questions require telescope projects of which astronomers did not dare to dream only a short time ago. These programs will not be multibillion-dollar projects like the Great Observatories program that included Hubble, but leaner and more specialized efforts using new technologies to provide more and better capabilities for less money.

The Origins missions can be divided into three generations, exploring increasingly newer technologies. The Hubble Space Telescope and similar missions such as GRO, AXAF, and SIRTF, are understood as predecessors. The first generation, with launches between 2001 and 2007, consists of three missions, of which the Next Generation Space Telescope may be the most important. From a number of studies by NASA and industry, several concepts have

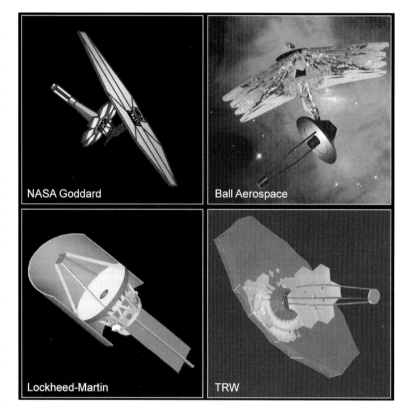

NASA Goddard

Ball Aerospace

Lockheed-Martin

TRW

The Next Generation Space Telescope may look like this. An internal NASA design complemented several independent studies by industry. The dominating feature is the giant solar panel that provides electrical power and also serves as a sun shade for the telescope and its electronics. The optical system (left) consists of the unfolded 8-meter main mirror and the secondary mirror on a truss. (NASA)

emerged. They all share the following basic requirements: NGST should have a main mirror with a diameter of at least 4 meters, and optimally 8 meters, to be unfolded in orbit; it should be sensitive in the wavelength range from 600 nanometers to 20 microns and optimized for 1 to 5 microns; and it should have

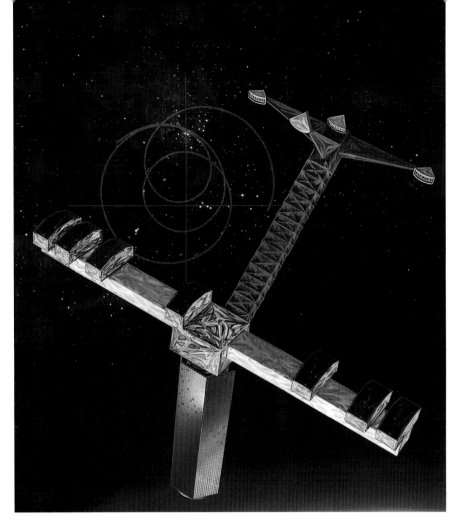

The satellite for NASA's Space Interferometry Mission may look like this. Several telescopes are mounted on a support structure; their light is combined coherently. Using optical interferometry, SIM will provide extremely accurate relative star positions and images with very high resolution of selected celestial objects. (NASA)

a resolution comparable to, or better than, Hubble. Finally, it should be able to observe objects that are 400 times fainter than those observable with modern large telescopes on the ground. NASA would like to begin the construction of the satellite around 2002 or 2003 — exactly at the time when the annual expenditures for Hubble will decrease significantly. The Space Telescope Science Institute in Baltimore would also begin to have free capacity at that time, and it could play an important role as the science and operations center for NGST. NGST could then be launched in 2007. The mission will cost only around half a billion dollars — truly cheap compared to the Hubble project. With its capability to observe young galaxies and to probe beyond redshifts of 5, NGST promises to become the main attraction of the Origins Program for at least five years.

The Space Interferometry Mission (SIM) is expected to be the second leg of the first generation of Origins missions. This optical interferometer in space will be launched in 2004 or 2005. Seven small telescopes will be mounted on a structure about 10 meters in length, collecting light and focusing it in a single point with an accuracy that allows the waves of light to interact, providing spatial information of extreme resolution. The primary goal of this mission is astrometry with micro-arc-second precision — that is, the accurate determination of relative star positions — but this instrument will also be able to provide images of comparable resolution. In addition, SIM will be able to use destructive interference with so-called nulling to make stars disappear, allowing us to search for planets in their vicinity. SIM will also be launched in a "folded" configuration and then deployed in its final orbit. Its cost may be as low as $300 million.

SIM may not be the only interferometer in the Origins Program. A related experiment is foreseen by NASA as part of a technology-oriented program. New Millennium is a series of cost-effective missions geared primarily toward the exploration of new technologies under space conditions; science, while part of the program, is only a secondary goal. Other New Millennium flights will target asteroids, comets, and Mars, but the "Deep Space 3" mission will remain close to Earth. This mission, also called New Millennium Interferometer (NMI), may be launched at the end of 2001. It consists of three independent

A view of the universe from the Next Generation Space Telescope. A "deep field" for the future telescope is simulated by analogy to the Hubble Deep Field; only 1/64 of the anticipated full field of 4 by 4 arc minutes is shown. According to model calculations, NGST would see about 10,000 galaxies with redshifts higher than 5 in an exposure of 10 hours per color. This simulation uses real galaxy images distributed by computer over a realistic universe and then "observed" by an imaginary NGST. Several typical redshifts are marked on the right. (Myungshin & Stockman)

satellites, two of which collect light and focus it on the third. Is it possible to have three satellites fly in space independently, while maintaining their relative positions to within several nanometers over distances of almost a kilometer? The challenge is much larger than for SIM, where the optical elements are mounted to a common structure — but NMI is mainly a project with a view toward the future.

Many of the lessons learned from the three missions of the first Origins generation will be applied to the second and third generations. Each of these phases is represented by only one mission. The next giant step will be the Terrestrial Planet Finder (TPF), which may be launched around 2010. It will explore the brightest thousand stars within a distance of 40 light years, find their Earthlike planets if they exist, provide images of them, and deliver spectra of their atmospheres. This must be done using optical interferometry, to obtain the necessary resolution. The basic design of TPF has not even been set yet — not until the New Millennium Interferometer experiment will it be known whether free-flying telescopes can be controlled with the required accuracy or must be mounted on a common structure. A conservative design foresees four to six 1.5-meter-diameter telescopes mounted on a structure about 70 to 75 meters in length; the relative positions of the telescopes would be controlled by laser interferometry to within a fraction of the wavelength of light. TPF would work at wavelengths of 7 to 17 microns, where planets appear particularly bright relative to stars.

Sometime "beyond the current horizon," as NASA prefers to call it, the third generation of the Origins Program would provide the crowning achievement: the Planet Imager (PI) should provide the first resolved images of the surfaces of planets around other stars. The previous generation, the Terrestrial Planet Finder, will be able to show planets only as faint points of light next to their suns. Obtaining images that show details of these distant planets is a giant leap. The PI, as currently envisioned, would require about five free-flying interferometers placed about 5000 kilometers apart. Each of them would contain several 8-meter telescopes (the maximum size of NGST!). Each of these satellites would remove the light from the bright sun of the distant solar system using the nulling method. The remaining light would be relayed to the central satellite, where it would interact again to produce the superresolution images. Today, this is science fiction — but after the first step of NGST, through further steps of NMI and SIM to TPF, the last step may be possible as well.

A vision for the future. Such
a constellation of giant
optical interferometers —
each carrying five 8-meter
telescopes — could be capable
of resolving surface details
of planets similar to Earth
around other stars. This
Planet Imager could become
reality during the first half
of the twenty-first century.
The next step would then be
a mission to one of these
promising planets. (JPL)

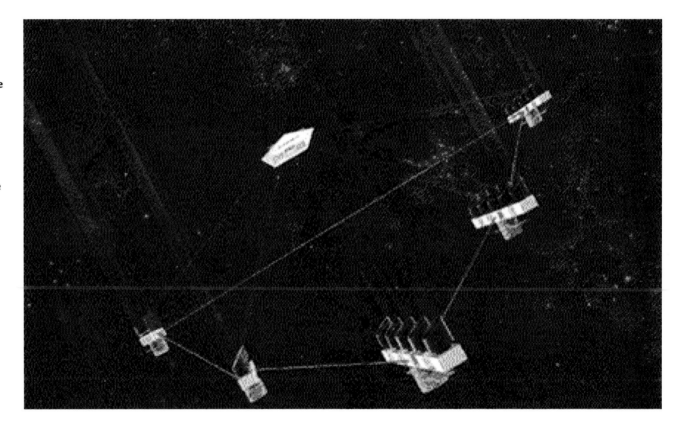

Europe as Partner

For several years, it appeared that the brave new world of space astronomy was reserved for the United States. After its final but expensive participation in the manned International Space Station project, the European Space Agency has let its science budget effectively shrink since 1995. The hopes of European astronomers for a new ESA instrument on the Hubble Space Telescope were shattered, and ESA could even have been completely eliminated from the Hubble program in 2001, when the memorandum of understanding between NASA and ESA runs out. The memorandum guaranteed European astronomers at least 15 percent of observing time on Hubble, in exchange for personnel resources and technical contributions, such as instruments and solar panels. Without a new instrument, the situation looked bleak, but the outlook began to improve in 1998: Hubble and the Next Generation Space Telescope could be treated as a package, assuring European astronomers access to both telescopes in exchange for contributions to NGST.

A substantial contribution to NGST could be financed by ESA under its new long-term science plan, reformed in 1997: the NGST contribution could become one of ESA's Flexi-Missions, with a budget of about $170–$200 million. Earlier, ESA's long-range science program, called Horizon 2000, contained large "cornerstone" missions and a set of still relatively expensive medium-sized projects. But that program turned out to be too inflexible, given the shrinking budget. The cornerstone missions will likely remain, but two new Flexi-Missions (or F-Missions) now replace the old medium-sized programs. The first F-Mission, or F1, would be a mission to Mars in 2003; F2, with a target date of 2007, would fit exactly into the schedule for NGST. Thus, Europe could continue to play an important role in this project. Given the available budget for an F-type mission, ESA could contribute as much as a third of the project cost of NGST by providing scientists, instruments, and satellite subsystems.

There are a number of competing projects for the ESA F2 mission from various areas of space science, however, and it will be incumbent on European astronomers to demand with one voice that ESA join the NGST project. In addition, there are other areas of overlap with NASA plans. Earlier versions of ESA's long-range plans also foresaw a large optical interferometer in space. According to the Horizon 2000 Plus plan, a relative of SIM, called the Global Astrometric Interferometer for Astrophysics (GAIA), could be launched around 2014; it would perform astrometry with extreme angular resolution. Alternatively, a relative of the Terrestrial Planet Finder, called Darwin, could provide infrared images with five 1-meter telescopes at baselines of 50 meters.

What could be more obvious than to coordinate the U.S. and European plans, to pool resources, and to open the next windows on the universe together?

Part 6

Appendix

Glossary

Accretion Disk: a disk of gas around a compact object, such as a star or a black hole, resulting from infalling matter. The material in the disk spirals around the central compact object, until it is finally accreted by it. Accretion disks and the associated central objects come in a wide range of sizes, from small stars to supermassive black holes in the centers of galaxies.

Andromeda Galaxy or Andromeda Nebula: the closest galaxy comparable to the Milky Way in size, also known as Messier 31 (M31). Its distance is 2.3 million light years. The two galaxies M31 and the Milky Way are the most massive members of the so-called Local Group of galaxies.

Antiparticle: an elementary particle composed of antimatter, with almost identical characteristics to those of normal matter. The main difference lies in the electrical charge: antiparticles have the opposite sign compared to normal elementary particles.

Asteroid or Planetoid or Small Planet: an irregularly shaped celestial object, consisting of various kinds or rocks, and with diameters from a few meters to 1000 kilometers. Most asteroids circle the Sun in a main belt between the orbits of Mars and Jupiter, but sometimes asteroids can come fairly close to Earth.

Atomic Nucleus: a collection of protons and neutrons held together by the strong force. The electrical charge of a nucleus is equal to the sum of the charges of the protons. The nucleus is about 100,000 times smaller than the atom itself. An atom has a diameter of about 10^{-12} meters.

Background Radiation: microwave radiation (in the short-wavelength radio domain) filling the entire universe. It was emitted when the universe was only about 300,000 years old. The temperature of this background radiation is about 2.7 Kelvin (or 2.7 degrees above absolute zero). The COBE satellite found that temperature variations from one point of the sky to another amount to less than 30 billionth Kelvin.

Big Bang: a cosmological concept, according to which the initially very hot and dense universe began in a giant explosion about 15 billion years ago. This explosion marks the beginning of the expansion of the universe, which still continues today.

Black Dwarf: the end product of the evolution of a star. Once a white dwarf has released all of its energy as radiation into space, it becomes an invisible stellar remnant. The fraction of dark matter made up of these black dwarfs is unclear.

Black Hole: the result of a collapse of a star with more than five solar masses. Such an object has a strong gravitational field combined with small size, to that neither light nor matter can escape from the black hole.

Comet: a roughly spherical body several kilometers in diameter and composed of ice and dust, traveling around the Sun in a highly elliptical orbit. With large telescopes the comet core can be observed up to very large distances from the Sun. When the core approaches the Sun, its ice partly evaporates and forms a huge tail, pointing away from the Sun as a result of the solar wind. The tail can reach a length of several hundred million kilometers.

Cosmic Radiation: particle radiation consisting of mainly protons and electrons. The particles have been accelerated to high energies by supernova explosions, interstellar magnetic fields, or processes in the vicinity of neutron stars.

Cosmology: the science of the formation and development of the universe and of its large-scale structures.

Critical Density: the density of matter at which the universe is flat or does not exhibit any curvature of space. In today's universe the critical density corresponds to an average of three hydrogen atoms per cubic meter. The expansion of a universe with critical density would end only after an infinitely long time. A universe at higher than critical density would begin to contract again at some time. Observations indicate that our universe is open: it has less then critical density and its expansion will continue forever.

Dark Matter: matter of unknown nature that does not emit radiation. The existence of this invisible matter has been deduced at first from studies of the motions of galaxies in clusters of galaxies, and from those of stars and gas in galaxies, and later also from the relative abundances of chemical elements formed shortly after the Big Bang.

Density Fluctuation: spatial variation in the distribution of matter in the universe. These fluctuations serve as seeds for the formation of galaxies, clusters of galaxies, and larger structures, particularly in the early universe. Density fluctuations can be inferred from minute variations of the temperature of the background radiation: these temperature variations amount to about 30 billionth of a degree.

Dwarf Galaxy: a galaxy of small size and mass. The average diameter of a dwarf galaxy is 15,000 light years, about six times smaller than a normal galaxy. Its mass ranges from 100 million to one billion solar masses, or about 1000 to 10,000 times less than the mass of a normal galaxy. Dwarf galaxies can have elliptical or irregular shape.

Electromagnetic Force: a force that only affects electrically charged particles. It causes particles of opposite electrical charge to attract, and particles of identical charge to repel each other.

Electron: the lightest elementary particle that carries electrical charge. The electron has a mass of 9×10^{-31} kilograms and carries a negative charge.

Elementary Particle: the basic building blocks of matter and radiation. The meaning of "elementary" has changed over time. Earlier, protons and neutrons were seen as elementary particles, but today we know that they are composed of quarks. Electrons, neutrinos, and photons are examples of truly elementary particles.

Elliptical Galaxy: a galaxy that appears as an ellipse in its projection to the sky, and in general does not exhibit any structure, such as spiral arms. Elliptical galaxies come as dwarfs or giants. Normally, they consist of old stars and only a small fraction of gas and dust.

Galactic Disk: a collection of stars, gas, and dust, in the form of a flattened disk in a spiral galaxy. The disk of our Milky Way has a diameter of about 90,000 light years, and a thickness of 3000 light years.

Galactic Halo: a spherically-shaped collection of old stars and globular clusters, surrounding a spiral galaxy. Observations indicate that the visible halo is surrounded by an invisible halo that is ten times larger and also contains ten times more mass.

Galaxy: a collection of about 10 million (in the case of a dwarf galaxy) to 10 trillion stars (in the case of a giant galaxy), bound together by their gravity. Galaxies are the building blocks of structures in the universe. A medium-size galaxy such as the Milky Way contains about 100 billion stars.

Galaxy Cannibalism: a process in which the motion of a galaxy is influenced by the gravitational pull from another more massive galaxy, resulting in the former falling into the latter. The devoured galaxy looses its identity, and its stars are mixed with those of the cannibalistic galaxy.

Galaxy Cluster: a dense collection of several thousand galaxies, bound together by their gravitation. The diameter of a typical galaxy cluster is 60 million light years, and its mass can reach a few hundred trillion solar masses.

Galaxy Group: a collection of typically 20 galaxies, bound together by their gravitation. The extent of such a group amounts to about six million light years, and its mass ranges from one to ten trillion solar masses.

Galaxy Supercluster: a collection of tens of thousands of galaxies located in groups and clusters, and bound together by their gravitation. Superclusters have a flattened shape, an average diameter of 90 million light years, and typically contain about one quadrillion solar masses.

Galaxy with Active Nucleus: a galaxy, in which the majority of radiation comes from a central region, a comparatively small nucleus with a diameter of only several light hours to light months, or about one billionth of the diameter of the galaxy. The energy released as radiation may be caused by a massive black hole that devours stars and gas in its vicinity.

Gamma Radiation: electromagnetic radiation consisting of photons of the highest energy.

Globular Cluster: a spherical collection of typically 100,000 stars, bound together by their gravity. The globular clusters of the Milky Way are very old objects.

Gravitation: an attracting force that affects all matter containing mass. It is the weakest of all forces in nature, but exhibits practically infinite reach.

Great Void in the Universe: a region of the universe containing almost no galaxies. Such regions may have diameters of several dozen million light years.

Heavy Elements: chemical elements heavier than helium. In astronomy, they are sometimes referred to as "metals". Heavy elements are produced by nuclear fusion in the interior of stars.

Helium: a chemical element with a nucleus consisting of two protons and two neutrons. Helium constitutes about one quarter of the matter in the universe. The majority of helium formed during the first three minutes after the Big Bang.

Hydrogen: the lightest chemical element. A hydrogen atom consists of a proton that is circled by an electron. Three quarters of the visible matter of the universe consist of hydrogen.

Interstellar Dust: small dust grains with a size of a millionth of a centimeter, generated in the outer layers of red giant stars. Interstellar dust predominantly absorbs blue light from stars and therefore makes them appear redder.

Irregular Galaxy: generally a dwarf galaxy that exhibits neither spiral structure nor elliptical shape. Irregular galaxies contain mostly young stars, gas, and dust.

Jet: focused streams of matter, originating in the cores of active galaxies, and usually appearing as pairs streaming in opposite directions from the active nucleus. Jets partly consist of fast electrons that interact with the magnetic field of the galaxy, causing radio emission from two large lobes surrounding the galaxy. On a smaller scale, jets are also seen in young or just forming stars, and in close binary systems.

Light Year: the distance traveled by a ray of light over one year. A light year is a distance of 9.46 trillion kilometers, or roughly 63,000 times the distance from Sun to Earth. Correspondingly, a light second equals 300,000 kilometers, a light minute 18 million kilometers, a light hour 1.1 billion kilometers, and a light month 788 billion kilometers.

Local Group: a group of galaxies that includes the Andromeda Galaxy and the Milky Way. Each of these two has a mass of about 1 trillion solar masses. Together, they dominate the local group. The other members of the local group, for instance the Magellanic Clouds, are dwarf galaxies of 10 million to 10 billion solar masses.

Local Supercluster: the supercluster of galaxies that includes the Milky Way. The local group of galaxies is located at the edge of the flattened disk of the local supercluster. The Virgo cluster forms the center of the local supercluster, hence it is also known as the Virgo supercluster.

Milky Way: a system of hundred billion stars, including our Sun. In contrast to other galaxies we can only see the Milky Way from the inside: it extends as a luminous band across the sky, and all stars visible by eye belong to it.

Neutrino: a neutral elementary particle, affected only by the weak force and to a small degree by gravitation; a mass has only be confirmed for one type of neutrino, the so-called muon neutrino. Neutrinos were formed in large numbers during the first moments of the universe; today, they still originate in the interior of stars and in supernova explosions.

Neutron: a neutral elementary particle consisting of three quarks. A neutron is 1838 times more massive than an electron, but only a little heavier than a proton. Atomic nuclei are made up of neutrons and protons.

Neutron Star: a celestial object with a typical diameter of 20 kilometers and a density of 10^{17} kilograms per cubic meter. Such a star contains between 1.4 and 5 solar masses; it has exhausted its nuclear fuel long ago.

Photon: this elementary particle of the electromagnetic radiation has no mass and travels at the speed of light (about 300,000 kilometers per second). Depending on the energy it carries, this particle can be observed (in descending order of energy) as gamma ray, as x ray, as an ultraviolet photon, as visible light, as an infrared photon, or as a radio wave.

Planet: a celestial object of more than about 1000 kilometers diameter that does not have any substantial source of energy of its own, and that orbits a star and reflects its light. In our own solar system nine planets are known; during recent years planets around a number of stars other than the Sun have been discovered.

Planetary Nebula: a gas shell ejected by a star, before that star developed from a red giant to a white dwarf. The star in the center of the shell causes it to glow. Viewed through a small telescope, a planetary nebula appears as a round disk, reminiscent in size and color of a distant planet like Uranus or Neptune – hence its (misleading) name.

Proton: an elementary particle with positive electrical charge consisting of three quarks. A proton is 1836 times more massive than an electron. Atomic nuclei are made up of neutrons and protons.

Quark: an elementary particle that is the constituent of neutrons and protons. Quarks have fractional electrical charges of 1/3 or 2/3 of the elementary charge of an electron. Quarks are affected by the strong force.

Quasar: a celestial object of star-like appearance, hence its name, derived from "quasi-stellar source". The light from quasars is shifted towards the red, indicating very large distances and correspondingly very high luminosities. Quasars are the most distant and brightest known objects in the universe.

Red Giant: a star that has used up hydrogen in its core as nuclear fuel and is now transforming helium into carbon and oxygen. In this phase the outer parts of the star expand to many times the diameter of the star at the beginning of its life, giving rise to the term giant. Red giants have comparatively cool surface temperatures and therefore appear red.

Spiral Galaxy: a galaxy with a spherical collection of stars in the center, the so-called bulge, surrounded by a flattened disk of stars, gas, and dust. Luminous young stars form conspicuous spiral arms in the disk.

Star: a sphere of gas, typically consisting of 75 percent hydrogen, 23 percent helium, and 2 percent heavier elements. Two forces balance such a sphere of gas: gravitation, trying to compress the star, and gas and radiation pressure caused by nuclear reactions in the interior of the star, attempting to expand the star.

Strong Force: the strongest of the four forces of nature. The strong force binds quarks to each other, forming neutrons and protons, and also binds neutrons and protons into atomic nuclei. It has an effect only within atomic nuclei (with diameters of 3×10^{-15} meters), and does not influence photons or electrons.

Weak Force: a force responsible for the decay of atoms and for radioactivity. It has an effect only over scales smaller than the diameter of an atom (about 10^{-17} meters).

White Dwarf: a small celestial object with a diameter of about 10,000 kilometers (comparable to the size of Earth), high density of 10^8 to 10^{11} kilograms per cubic meter, and not more than 1.4 solar masses. A white dwarf is the result of the evolution of a low-mass star that has exhausted its nuclear fuel.

X Rays: electromagnetic radiation consisting of photons of very high energy.

Want to See More? (World Wide Web Addresses)

Latest News and Pictures from the Hubble Space Telescope

http://www.stsci.edu/

News from NASA

http://www.nasa.gov/
http://www.nasa.gov/today/index.html

News from ESA

http://www.esrin.esa.it/htdocs/esa/esa.html

General Astronomical Pictures

http://antwrp.gsfc.nasa.gov/apod

Solar System and Planets

http://www.jpl.nasa.gov/
http://photojournal.jpl.nasa.gov/
http://seds.lpl.arizona.edu/nineplanets/nineplanets/
  nineplanets.html
http://hplyot.obspm.fr/np/nineplanets/
  nineplanets.html
http://bang.lanl.gov/solarsys/

Planets around Other Stars

http://cannon.sfsu.edu/~williams/planetsearch/

planetsearch.html

Extrasolar Visions

http://www.jtwinc.com/Extrasolar/evwarn.html

Other Worlds, Distant Suns

http://garber.simplenet.com/main.htm

Search for Extraterrestrial Intelligence (SETI)

http://www.seti-inst.edu/

Nebulae, Star Clusters, and Galaxies

http://zebu.uoregon.edu/messier.html
http://crux.astr.ua.edu/choosepic.html

Cosmology

http://www.astro.ubc.ca/people/scott/cosmology.html

General Links of Astronomical Interest

http://www.stsci.edu/science/net-resources.html
http://cdsweb.u-strasbg.fr/astroweb.html
http://www.cv.nrao.edu/fits/www/astronomy.html
http://hea-www.harvard.edu/QEDT/jcm/space/
  space.html

Further Reading

History of Astronomy and General Introductions

Jean Audouze and Guy Israel, eds. *The Cambridge Atlas of Astronomy.* Third Edition. Cambridge: Cambridge University Press, 1994.

William K. Hartmann. *The Cosmic Voyage: Through Time and Space.* Belmont, Calif.: Wadsworth, 1992.

Nigel Henbest and Michael Marten. *The New Astronomy.* Second Edition. Cambridge: Cambridge University Press, 1996.

William J. Kaufmann III and Neil F. Comins. *Universe.* Fourth Edition. New York: W. H. Freeman, 1993.

David Leverington. *A History of Astronomy: From 1890 to the Present.* London: Springer-Verlag, 1996.

John North. *The Norton History of Astronomy and Cosmology.* New York: W. W. Norton, 1994.

Archie Roy, ed. *Oxford Illustrated Encyclopedia of the Universe.* Oxford: Oxford University Press, 1992.

Hugh Thurston. *Early Astronomy.* New York: Springer-Verlag, 1994.

The Hubble Space Telescope

Eric Chaisson. *The Hubble Wars: Astrophysics Meets Astropolitics in the Two-Billion-Dollar Struggle over the Hubble Space Telescope.* New York: Harper Collins, 1994.

Carolyn Collins Peterson and John C. Brandt. *Hubble Vision: Astronomy with the Hubble Space Telescope.* Cambridge: Cambridge University Press, 1995.

George Field and Donald Goldsmith. *The Space Telescope: Eyes Above the Atmosphere.* Chicago: Contemporary Books, 1989.

Daniel Fischer and Hilmar Duerbeck. *Hubble: A New Window to the Universe.* New York: Springer-Verlag, 1996.

Robert Smith. *The Space Telescope: A Study of NASA, Science, Technology, and Politics.* Cambridge: Cambridge University Press, 1989.

Edwin Hubble

Gale E. Christianson. *Edwin Hubble: Mariner of the Nebulae.* New York: Farrar, Straus and Giroux, 1995.

Edwin Hubble. *The Realm of the Nebulae.* New Haven, Conn.: Yale University Press, 1983.

Alexander S. Sharov and Igor D. Novikov. *Edwin Hubble, the Discoverer of the Big Bang Universe.* Cambridge: Cambridge University Press, 1993.

Cosmology

John Barrow. *The Origin of the Universe.* New York: Basic Books, 1994.

Alan Michael Dressler. *Voyage to the Great Attractor: Exploring Intergalactic Space.* New York: Knopf, 1994.

Albert Einstein. *The Meaning of Relativity.* Fifth

Edition. Princeton, N.J.: Princeton University Press, 1956.

Alan H. Guth and Alan Lightman. *The Inflationary Universe. The Quest for a New Theory of Cosmic Origins.* Reading, Mass.: Addison-Wesley, 1997.

Stephen Hawking. *A Brief History of Time.* New York: Bantam, 1988.

Rudolph Kippenhahn. *Light from the Depths of Time.* New York: Springer-Verlag, 1987.

Rocky Kolb. *Blind Watchers of the Sky: The People and Ideas that Shaped Our View of the Universe.* Reading, Mass.: Addison-Wesley, 1996.

John C. Mather and John Boslogh. *The Very First Light: The True Inside Story of the Scientific Journey Back to the Dawn of the Universe.* New York: Basic Books, 1996.

P. J. E. Peebles. *Principles of Physical Cosmology.* Princeton, N.J.: Princeton University Press, 1993.

Joseph Silk. *A Short History of the Universe.* New York: Scientific American Library, 1997.

George Smoot and Keay Davidson. *Wrinkles in Time.* New York: Morrow, 1994.

Kip S. Thorne. *Black Holes and Time Warps: Einstein's Outrageous Legacy.* New York: W. W. Norton, 1994.

Robert Wald. *Space, Time and Gravity: The Theory of the Big Bang and Black Holes.* Second Edition. Chicago: Chicago University Press, 1992.

Steven Weinberg. *The First Three Minutes: A Modern View of the Origin of the Universe.* New York: Basic Books, 1993.

Stars and Galaxies

Mitchell Begelman and Martin J. Rees. *Gravity's Fatal Attraction: Black Holes in the Universe.* New York: Scientific American Library, 1996.

Alan Dressler. *Voyage to the Great Attractor: Exploring Intergalactic Space.* New York: Knopf, 1994.

David Eicher, ed. *Deep-Sky Observing with Small Telescopes.* New York: Enslow, 1990.

Nigel Henbest and Heather Couper. *The Guide to the Galaxy.* Cambridge: Cambridge University Press, 1994.

James B. Kaler: *Cosmic Clouds: Birth, Death, and Recycling in the Galaxy.* New York: Scientific American Books, 1997.

James B. Kaler. *Stars and Their Spectra: An Introduction to the Spectral Sequence.* Cambridge: Cambridge University Press, 1998.

Donald E. Osterbrock: *Stars and Galaxies: Citizens of the Universe.* Readings from Scientific American Magazine. New York: W. H. Freeman, 1990.

Jean-Claude Pecker. *The Future of the Sun.* New York: McGraw-Hill, 1992.

Gareth Wynn-Williams. *The Fullness of Space: Nebulae, Stardust and the Interstellar Medium.* Cambridge: Cambridge University Press, 1991.

The Solar System

John C. Brandt and R. D. Chapman. *Rendezvous in Space: The Science of Comets.* New York: W. H. Freeman, 1992.

Nicholas Booth. *Exploring the Solar System.* Cambridge: Cambridge University Press, 1996.

Ronald Greeley and Raymond Batson. *The NASA Atlas of the Solar System.* Cambridge: Cambridge University Press, 1996.

Nigel Henbest. *The Planets: Portraits of New Worlds.* New York: Penguin USA, 1994.

Kenneth R. Lang and Charles A. Whitney. *Wanderers in Space: Exploration and Discovery in the Solar System.* Cambridge: Cambridge University Press, 1991.

David Morrison. *Exploring Planetary Worlds.* New York: Scientific American Library, 1993.

Planets Outside the Solar System

Ken Crosswell. *Planet Quest: The Epic Discovery of Alien Solar Systems.* New York: Free Press, 1997.

Donald Goldsmith. *Worlds Unnumbered.* Sausalito, Calif.: University Science Books, 1997.

Index

30 Doradus Nebula, 110
47 Tucanae, 118–119

Abell 2218, 85
Accretion Disk, 76, 203
Advanced Camera for Surveys
    (ACS), 8, 188, 189
Advanced X-Ray Astrophysics
    Facility (AXAF), 22, 88, 192
Akers, Tom, 31
Alpha Orionis, 112, 113
Andromeda Galaxy, 44, 203
Antenna Galaxy, 69–71
Arp, Halton, 68
Asteroids, 172–174, 203
Atlas (moon of Saturn), 161

Bahcall, John, 30
Becklin-Neugebauer object, 99–101
Beckwith, Steven V. W., 186–187
Beppo-SAX Satellite, 16, 22, 89
Beta Pictoris, 29, 151, 152
Betelgeuse, 112, 113
Big Bang, 50, 56, 203
Binary Stars, 145
Black Hole, 8, 72, 76, 79–81, 92,
    120, 126, 203
Blue Stragglers, 118–119
Bok Globules, 106

Bowersox, Kenneth, 33
Brown Dwarf, 151, 153–154
Butterfly Nebula, 134

Cannon, Annie, 141
Cartwheel Galaxy, 69
Cassini Space Probe, 163
Centaurus A, 69, 72–74
Cepheids, 44, 59, 60, 61
Charon, 155
Comet, 203
    Hale-Bopp, 175, 176
    Hyakutake, 176
Compton Gamma Ray
    Observatory (GRO), 16, 88,
    192
Cone Nebula, 103, 107
Corrective Optics Space Telescope
    Axial Replacement
    (COSTAR), 27–32, 189
Cosmic Background Explorer
    (COBE), 17, 54, 55
    Diffuse Infrared Background
       Experiment (DIRBE), 54
Cosmic Origins Spectrograph
    (COS), 189
Cosmological Constant, 43, 56,
    61, 62
Cosmology, 43–45, 204

Crab Nebula, 126
Critical Density of the Universe,
    56, 57, 62, 204

Dark Matter, 57, 204
Dwarf Novae, 145

Eagle Nebula (M16), 100, 102–103
Egg Nebula, 130, 131, 141
Einstein, Albert, 82
Einstein Cross, 29, 83
Einstein Ring, 82, 84
Enceladus (moon of Saturn), 163
Eta Carinae, 110, 115–117
European Southern Observatory
    (ESO), 12, 19, 102
    Very Large Telescope (VLT), 19,
       20
European Space Agency (ESA), 12,
    23, 186, 198
Evaporating Gaseous Globules
    (EGGs), 100, 102, 103

Fabricius, David, 146
Faint Object Camera (FOC), 23,
    27, 188, 189
Faint Object Spectrograph (FOS),

33, 34, 189
Far Infrared and Submillimeter
    Space Telescope (FIRST), 22
Fine Guidance Sensors (FGSs), 34,
    188
GK Persei, 146

Galaxies, 65–93
    active, 80–81, 205
    colliding, 68–72, 78
    formation of, 65–66
    mergers of, 65, 77
    starburst, 68, 86
Galaxy Classification, 44
Galaxy Clusters, 43, 205
Galaxy Superclusters, 43, 205
Galileo Space Probe, 166, 168,
    169, 171
Gamma Ray Bursts, 88–93
Ganymede (moon of Jupiter), 169,
    171
Gemini Project, 19
Giant Stars, 112–117
Gliese 106C, 154
Gliese 229B, 153–154
Global Astrometric Interferometer
    for Astrophysics (GAIA),
    198
Globular Clusters, 118, 119, 205

core collapse, 120
Goddard High Resolution
    Spectrograph (GHRS), 33,
    34, 189
Goddard Space Flight Center, 32,
    33
Gravitational Lenses, 29, 82–87

H-II Regions, 97
Hale, George Ellery, 15
Halley, Edmund, 115
Harbaugh, Gregory, 33, 34, 36
Hawley, Steven A., 8, 9, 11, 26, 33
Helix Nebula, 138, 139
Henize 1357, 128, 129
Hertzsprung-Russell Diagram, 118
High-Speed Photometer (HSP), 28
Hobby-Eberly Telescope (HET), 19
Hoffman, Jeff, 30, 31
Horowitz, Scott, 33
Hourglass Nebula, 106, 140, 141
Hubble Constant, 50, 56, 57, 59,
    60, 61, 86
Hubble Data Archive (HDA), 185,
    190
Hubble Deep Field (HDF), 11,
    46–55, 65, 72, 185, 194
Hubble Diagram, 56
Hubble Law, 45, 56

Hubble Space Telescope (HST)
    end of mission, 191
    observing efficiency, 185
    optical flaw, 27
    solar panels, 30, 34, 186, 188
Hubble, Edwin P., 25, 30, 44, 45,
    52, 56, 60
Hypernovae, 92

Inflationary Universe, 57
Infrared Astronomy Satellite
    (IRAS), 17
Infrared Space Observatory (ISO),
    16, 17, 21, 103
Interstellar Matter, 97
Io (moon of Jupiter), 169, 170

Jupiter, 166–168

Keck Observatory, 17, 18, 53
Kuiper Belt, 155–156

Lagoon Nebula (M8), 103, 108
Large Magellanic Cloud, 29, 110,
    121
Large Space Telescope (LST), 23

Lee, Mark, 33, 35, 36
Lyman Limit, 48

M2–9, 134
M4, 118
M15, 119
M33, 111
M87, 75
M100, 32
M105, 80
Magellan Project, 19
Magellanic Clouds, 110
Mariner 9, 178
Mars, 177–181
Mars Global Surveyor, 178–179
Mars Pathfinder, 177–181
Mayall, Margaret, 141
Mikulski, Barbara, 32
Mira, 8, 147
Musgrave, Story, 30, 31
MyCn 18, 140, 141

Near-Infrared Camera and
    Multi-Object Spectrometer
    (NICMOS), 8, 33, 38–40, 55
Neptune, 157–158
Neutron Stars, 91, 92, 126–127,
    147, 207

New Millenium Interferometer
    (NMI), 193, 196
Next Generation Space Telescope
    (NGST), 9, 22, 66, 186, 192,
    194, 196, 198
NGC 604, 111
NGC 1667, 81
NGC 2363, 110
NGC 2366, 110
NGC 3377, 80
NGC 3379, 80
NGC 3982, 81
NGC 4038, 69
NGC 4039, 69
NGC 4151, 80
NGC 4478, 75
NGC 4486B, 75, 80
NGC 5307, 135
NGC 6251, 81
NGC 7027, 136
NGC 7293, 138, 139
Nicollier, Claude, 30
Nova Cygni 1992, 145
Novae, 145–146

Oberth, Hermann, 23
Olbers, Wilhelm, 46
Olbers' Paradox, 46
Origins Program of NASA,

192–197
Orion Nebula, 97–102, 110, 151

P Cygni, 110
Pistol Nebula, 113–114
Pistol Star, 112–114
Planet Formation, 99, 151
Planet Imager (PI), 196
Planetary Nebulae, 128–144, 147,
    207
Planets, 151–181
    around other stars, 151–154
Pluto, 155, 156
Poe, Edgar Allen, 50
Prometheus (moon of Saturn), 161
Protoplanetary Disks (Proplyds),
    97, 98, 102
Proxima Centauri, 153
Pulsars, 126–127

Quasars, 76–79, 81, 83, 207
    central engine, 76
    host galaxies, 77

R Leonis, 147
Red Giants, 146, 207
Redshift, 45, 48, 76

216

Roentgen Satellite (ROSAT), 22,
    122, 127

Saturn, 161–165
Servicing Mission 1, 8, 30–32
Servicing Mission 2, 8, 33–37
Servicing Mission 3, 36, 40, 188
Servicing Mission 4, 188–189
Seyfert, Carl, 80
Seyfert Galaxies, 80–81
Smith, Steven, 33, 35, 36
South Atlantic Anomaly (SAA), 38
Space Infrared Telescope Facility
    (SIRTF), 21, 88, 192
Space Interferometry Mission
    (SIM), 193, 196
Space Shuttle
    Atlantis, 88
    Challenger, 25
    Discovery, 7, 25–27, 33, 35–37
    Endeavor, 32
Space Telescope European
    Coordinating Facility, 25,
    190
Space Telescope Imaging
    Spectrograph (STIS), 8,

38–40, 55, 189
Space Telescope Science Institute
    (STScI), 12, 23, 186, 187,
    190
Speed of Light, 43
Spitzer, Lyman, 23, 24
Star Formation, 97, 102, 103, 110
Star-Forming Regions, 110–111
Stellar Evolution, 118, 128
Stingray Nebula, 128
Subaru Telescope, 17, 18
Supernova 1987A, 29, 121–124
Supernova Remnants, 126–127
Supernovae, 121
    as distance indicators, 61–63

T Pyxidis, 145, 146
Tanner, Joseph, 33, 34, 36
Terrestrial Planet Finder (TPI), 196
Thornton, Kathy, 31
Toomre, Alar, 68
Toomre, Juri, 68
Trans-Neptunian Objects, 155–56

Uranus, 158–160

Vesta, 172, 173
Viking Space Probes, 177, 178
Virgo Cluster, 59, 72, 75
Voyager Space Probes, 157, 158,
    161, 169

W Hydrae, 147
White Dwarfs, 126, 145, 146, 208
Wide-Field and Planetary Camera
    (WFPC), 26, 30, 31, 189
Wide-Field and Planetary Camera
    2 (WFPC2), 30, 31, 40, 55,
    188, 189
Wide-Field Camera 3 (WFC3), 189
Williams, Robert E., 11, 186

X-Ray Multi-Mirror Mission
    (XMM), 22

Young Stellar Objects (YSOs), 102

Zwicky, Fritz, 68, 82